ANNUAL EDITIONS

Geography

Twentieth Edition

EDITOR

Gerald R. Pitzl

Macalester College (retired)

Gerald R. Pitzl received his bachelor's degree in secondary social science education from the University of Minnesota in 1964 and his M.A. (1971) and Ph.D. (1974) in geography from the same institution. Dr. Pitzl has taught a wide array of geography courses, and he is the author of a number of articles on geography, the developing world, and the use of the Harvard Case Method. His book, *Encyclopedia of Human Geography,* was published in January 2004. This one-volume work is designed for use in the Advanced Placement (AP) Program in Human Geography. Dr. Pitzl continues to conduct workshops and in-service sessions on active learning and case-based discussion methods.

McGraw-Hill/Dushkin
2460 Kerper Blvd., Dubuque, IA 52001

Visit us on the Internet
http://www.dushkin.com

Credits

1. **Geography in a Changing World**
   Unit photo—© by PhotoDisc, Inc.
2. **Human-Environment Relationships**
   Unit photo—© Getty Images/PhotoLink/C. Sherburne
3. **The Region**
   Unit photo—© Getty Images/Ryan McVay
4. **Spatial Interaction and Mapping**
   Unit photo— © CORBIS/Royalty-Free
5. **Population, Resources, and Socioeconomic Development**
   Unit photo—© Getty Images/Punchstock

Copyright

Cataloging in Publication Data
Main entry under title: Annual Editions: Geography. 2005/2006.
1. Geography—Periodicals. I. Pitzl, Gerald R., *comp.* II. Title: Geography.
ISBN 0–07–310838–3 658'.05 ISSN 1091–9937

Twentieth Edition

Cover image © Photos.com and Harnett/Hanson/Getty Images
Printed in the United States of America 1234567890QPDQPD987654 Printed on Recycled Paper

Editors/Advisory Board

Members of the Advisory Board are instrumental in the final selection of articles for each edition of ANNUAL EDITIONS. Their review of articles for content, level, currentness, and appropriateness provides critical direction to the editor and staff. We think that you will find their careful consideration well reflected in this volume.

Staff

Preface

In publishing ANNUAL EDITIONS we recognize the enormous role played by the magazines, newspapers, and journals of the public press in providing current, first-rate educational information in a broad spectrum of interest areas. Many of these articles are appropriate for students, researchers, and professionals seeking accurate, current material to help bridge the gap between principles and theories and the real world. These articles, however, become more useful for study when those of lasting value are carefully collected, organized, and reproduced in a low-cost format, which provides easy and permanent access when the material is needed. That is the role played by ANNUAL EDITIONS.

The articles in this twentieth edition of Annual Editions: Geography represent the wide range of topics associated with the discipline of geography. The major themes of spatial relationships, regional development, the population explosion, and socioeconomic inequalities exemplify the diversity of research areas within geography.

The book is organized into five units, each containing articles relating to geographical themes. Selections address the conceptual nature of geography and the global and regional problems in the world today. This latter theme reflects the geographer's concern with finding solutions to these serious issues. Regional problems, such as food shortages in the Sahel and the greenhouse effect, concern not only geographers but also researchers from other disciplines.

The association of geography with other fields is important, because expertise from related research will be necessary in finding solutions to some difficult problems. Input from the focus of geography is vital in our common search for solutions. This discipline has always been integrative. That is, geography uses evidence from many sources to answer the basic questions, "Where is it?" "Why is it there?" and "What is its relevance?" The first group of articles emphasizes the interconnectedness not only of places and regions in the world but of efforts toward solutions to problems as well. No single discipline can have all of the answers to the problems facing us today; the complexity of the issues is simply too great.

The writings in unit 1 discuss particular aspects of geography as a discipline and provide examples of the topics presented in the remaining four sections. Units 2, 3, and 4 represent major themes in geography. Unit 5 addresses important problems faced by geographers and others.

Annual Editions: Geography 05/06 will be useful to both faculty members and students in their study of geography. The anthology is designed to provide detail and case study material to supplement the standard textbook treatment of geography. The goals of this anthology are to introduce students to the richness and diversity of topics relating to places and regions on the Earth's surface, to pay heed to the serious problems facing humankind, and to stimulate the search for more information on topics of interest. As such, this anthology is an ideal companion volume for use in the Advanced Placement (AP) Program in Human Geography found in high schools across the country. This program, like others sponsored by The College Board, has grown steadily since its inception. Currently, over 9,000 high school students are enrolled in the AP Human Geography Program.

I would like to express my gratitude to Devi Benjamin for her encouragement and invaluable assistance in preparing the nineteenth edition. Without her enthusiasm and professional efforts, this project would not have moved along as efficiently as it did. Special thanks are also extended to the McGraw-Hill/Dushkin editorial staff for coordinating the production of the reader. A word of thanks must go as well to all those who recommended articles for inclusion in this volume and who commented on its overall organization.

In order to improve the next edition of Annual Editions: Geography, we need your help. Please share your opinions by filling out and returning to us the postage-paid ARTICLE RATING FORM found on the last page of this book. We will give serious consideration to all your comments and recommendations.

Gerald R. Pitzl
Editor

Contents

UNIT 1
Geography in a Changing World

The concepts in bold italics are developed in the article. For further expansion, please refer to the Topic Guide.

UNIT 2
Human-Environment Relations

The concepts in bold italics are developed in the article. For further expansion, please refer to the Topic Guide.

UNIT 3
The Region

The concepts in bold italics are developed in the article. For further expansion, please refer to the Topic Guide.

UNIT 4
Spatial Interaction and Mapping

The concepts in bold italics are developed in the article. For further expansion, please refer to the Topic Guide.

UNIT 5
Population, Resources, and Socioeconomic Development

The concepts in bold italics are developed in the article. For further expansion, please refer to the Topic Guide.

The concepts in bold italics are developed in the article. For further expansion, please refer to the Topic Guide.

Topic Guide

This topic guide suggests how the selections in this book relate to the subjects covered in your course. You may want to use the topics listed on these pages to search the Web more easily.

On the following pages a number of Web sites have been gathered specifically for this book. They are arranged to reflect the units of this *Annual Edition*. You can link to these sites by going to the DUSHKIN ONLINE support site at *http://www.dushkin.com/online/*.

ALL THE ARTICLES THAT RELATE TO EACH TOPIC ARE LISTED BELOW THE BOLD-FACED TERM.

World Wide Web Sites

The following World Wide Web sites have been carefully researched and selected to support the articles found in this reader. The easiest way to access these selected sites is to go to our DUSHKIN ONLINE support site at *http://www.dushkin.com/online/*.

AE: Geography 05/06

The following sites were available at the time of publication. Visit our Web site—we update DUSHKIN ONLINE regularly to reflect any changes.

General Sources

About: Geography
http://geography.about.com
This Web site, created by the About network, contains hyperlinks to many specific areas of geography, including cartography, population, country facts, historic maps, physical geography, topographic maps, and many others.

The Association of American Geographers (AAG)
http://www.aag.org
Surf this site of the Association of American Geographers to learn about AAG projects and publications, careers in geography, and information about related organizations.

Geography Network
http://www.geographynetwork.com
The Geography Network is an online resource to discover and access geographical content, including live maps and data, from many of the world's leading providers.

National Geographic Society
http://www.nationalgeographic.com
This site provides links to National Geographic's huge archive of maps, articles, and other documents. Search the site for information about worldwide expeditions of interest to geographers.

The New York Times
http://www.nytimes.com
Browsing through the archives of the *New York Times* will provide you with a wide array of articles and information related to the different subfields of geography.

Social Science Internet Resources
http://www.wcsu.ctstateu.edu/library/ss_geography_cartography.html
This site is a definitive source for geography-related links to universities, browsers, cartography, associations, and discussion groups.

U.S. Geological Survey (USGS)
http://www.usgs.gov
This site and its many links are replete with information and resources for geographers, from explanations of El Niño, to mapping, to geography education, to water resources. No geographer's resource list would be complete without frequent mention of the USGS.

UNIT 1: Geography in a Changing World

Alternative Energy Institute (AEI)
http://www.altenergy.org
The AEI will continue to monitor the transition from today's energy forms to the future in a "surprising journey of twists and turns." This site is the beginning of an incredible journey.

Mission to Planet Earth
http://www.earth.nasa.gov
This site will direct you to information about NASA's Mission to Planet Earth program and its Science of the Earth System. Surf here to learn about satellites, El Niño, and even "strategic visions" of interest to geographers.

Poverty Mapping
http://www.povertymap.net
Poverty maps can quickly provide information on the spatial distribution of poverty. Here you will find maps, graphics, data, publications, news, and links that provide the public with poverty mapping from the global to the subnational level.

Santa Fe Institute
www.santafe.edu/index.php
This home page of the Santa Fe Institute—a nonprofit, multidisciplinary research and education center—will lead you to a plethora of valuable links related to its primary goal: to create a new kind of scientific research community, pursuing emerging science. Such links as Evolution of Language, Ecology, and Local Rules for Global Problems are offered.

Solstice: Documents and Databases
http://solstice.crest.org
In this online source for sustainable energy information, the Center for Renewable Energy and Sustainable Technology (CREST) offers information and databases on renewable energy, energy efficiency, and sustainable living. The site also offers related Web sites, case studies, and policy issues. Solstice also connects to CREST's Web presence.

UNIT 2: Human-Environment Relations

Alliance for Global Sustainability (AGS)
http://www.global-sustainability.org
The AGS is a cooperative venture seeking solutions to today's urgent and complex environmental problems. Research teams from four universities study large-scale, multidisciplinary environmental problems that are faced by the world's ecosystems, economies, and societies.

Human Geography
http://www.geog.le.ac.uk/cti/hum.html
The CTI Centre for Geography, Geology, and Meteorology provides this site, which contains links to human geography in relation to agriculture, anthropology, archaeology, development geography, economic geography, geography of gender, and many others.

The North-South Institute
http://www.nsi-ins.ca/ensi/index.html
Searching this site of the North-South Institute—which works to strengthen international development cooperation and enhance gender and social equity—will help you find information on a variety of development issues.

United Nations Environment Programme (UNEP)
http://www.unep.ch
Consult this home page of UNEP for links to critical topics of concern to geographers, including desertification and the impact

of trade on the environment. The site will direct you to useful databases and global resource information.

US Global Change Research Program

http://www.usgcrp.gov

This government program supports research on the interactions of natural and human-induced changes in the global environment and their implications for study. Find details on the atmosphere, climate change, global carbon and water cycles, ecosystems, and land use plus human contributions and responses.

World Health Organization

http://www.who.int

This home page of the World Health Organization will provide you with links to a wealth of statistical and analytical information about health in the developing world.

UNIT 3: The Region

AS at UVA Yellow Pages: Regional Studies

http://xroads.virginia.edu/~YP/regional.html

Those interested in American regional studies will find this site a gold mine. Links to periodicals and other informational resources about the Midwest/Central, Northeast, South, and West regions are provided here.

Can Cities Save the Future?

http://www.huduser.org/publications/econdev/habitat/prep2.html

This press release about the second session of the Preparatory Committee for Habitat II is an excellent discussion of the question of global urbanization.

IISDnet

www.iisd.org

The International Institute for Sustainable Development, a Canadian organization, presents information through gateways entitled Business and Sustainable Development, Developing Ideas, and Hot Topics. Linkages provide a multimedia resource for environment and development policy makers.

NewsPage

http://www.individual.com

Individual, Inc., maintains this business-oriented Web site. Geographers will find links to much valuable information about such fields as energy, environmental services, media and communications, and health care.

Treaty on Urbanization

http://www.geocities.com/atlas/urb/tretyurb.html

The original 1992 Treaty on Urbanization is available at this site. Its goal is just, democratic, and sustainable cities, town, and villages.

Virtual Seminar in Global Political Economy/Global Cities & Social Movements

http://csf.colorado.edu/gpe/gpe95b/resources.html

This Web site is rich in links to subjects of interest in regional studies, such as sustainable cities, megacities, and urban planning. Links to many international nongovernmental organizations are included.

World Regions & Nation States

http://www.worldcapitalforum.com/worregstat.html

This site provides strategic and competitive intelligence on regions and individual states, geopolitical analyses, geopolitical factors of globalization, geopolitics of production, and much more.

UNIT 4: Spatial Interaction and Mapping

Edinburgh Geographical Information Systems

http://www.geo.ed.ac.uk/home/gishome.html

This valuable site, hosted by the Department of Geography at the University of Edinburgh, provides information on all aspects of Geographic Information Systems and provides links to other servers worldwide. A GIS reference database as well as a major GIS bibliography is included.

Geography for GIS

http://www.ncgia.ucsb.edu/cctp/units/geog_for_GIS/GC_index.html

This hyperlinked table of contents was created by Robert Slobodian of Malaspina University. Here you will find information regarding GIS technology.

GIS Frequently Asked Questions and General Information

http://www.census.gov/geo/www/faq-index.html

Browse through this site to get answers to FAQs about Geographic Information Systems. It can direct you to general information about GIS as well as guidelines on such specific questions as how to order U.S. Geological Survey maps. Other sources of information are also noted.

International Map Trade Association

http://www.maptrade.org

The International Map Trade Association offers this site for those interested in information on maps, geography, and mapping technology. Lists of map retailers and publishers as well as upcoming IMTA conferences and trade shows are noted.

PSC Publications

http://www.psc.isr.umich.edu

Use this site and its links from the Population Studies Center of the University of Michigan for spatial patterns of immigration and discussion of white and black flight from high immigration metropolitan areas in the United States.

UNIT 5: Population, Resources, and Socioeconomic Development

African Studies WWW (U.Penn)

http://www.sas.upenn.edu/African_Studies/AS.html

Access to rich and varied resources that cover such topics as demographics, migration, family planning, and health and nutrition is available at this site.

Geography and Socioeconomic Development

http://www.ksg.harvard.edu/cid/andes/Documents/Background%20Papers/Geography&Socioeconomic%20Development.pdf

John L. Gallup wrote this 19-page background paper examining the state of the Andean region. He explains the strong and pervasive effects geography has on economic and social development.

Human Rights and Humanitarian Assistance

http://www.etown.edu/vl/humrts.html

Through this site, part of the World Wide Web Virtual Library, you can conduct research into a number of human-rights topics in order to gain a greater understanding of the issues affecting indigenous peoples in the modern era.

Hypertext and Ethnography

http://www.umanitoba.ca/faculties/arts/anthropology/tutor/aaa_presentation.new.html

This site, presented by Brian Schwimmer of the University of Manitoba, will be of great value to people who are interested in culture and communication. He addresses such topics as multivocality and complex symbolization, among many others.

Research and Reference (Library of Congress)

http://lcweb.loc.gov/rr/

This research and reference site of the Library of Congress will lead you to invaluable information on different countries. It provides links to numerous publications, bibliographies, and guides in area studies that can be of great help to geographers.

Space Research Institute

http://arc.iki.rssi.ru/eng/

Browse through this home page of Russia's Space Research Institute for information on its Environment Monitoring Information Systems, the IKI Satellite Situation Center, and its Data Archive.

World Population and Demographic Data

http://geography.about.com/cs/worldpopulation/

On this site you will find information about world population and additional demographic data for all the countries of the world.

We highly recommend that you review our Web site for expanded information and our other product lines. We are continually updating and adding links to our Web site in order to offer you the most usable and useful information that will support and expand the value of your Annual Editions. You can reach us at: *http://www.dushkin.com/annualeditions/*.

World Map

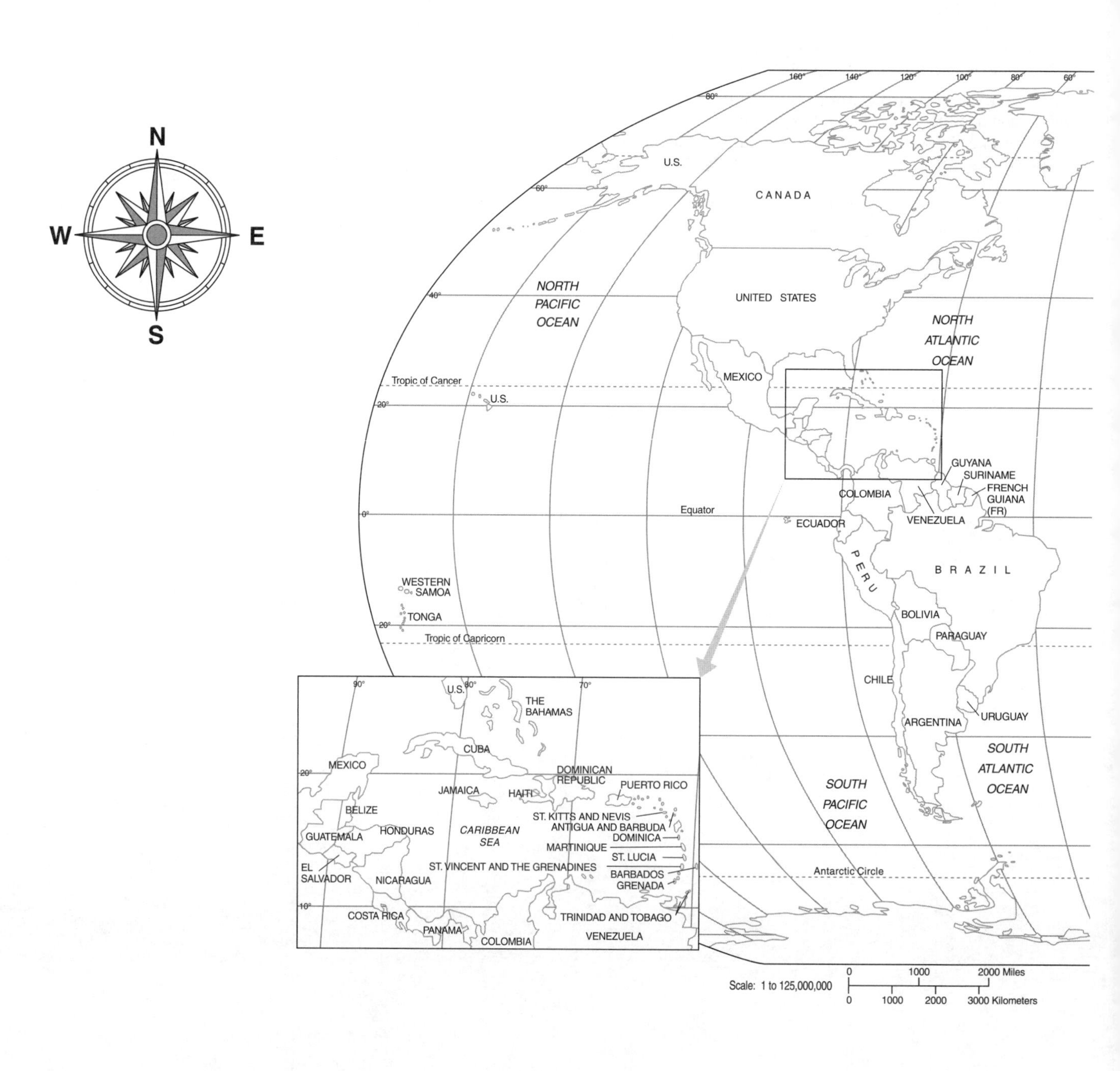

N
S
W
E
U.S.
CANADA
UNITED STATES
MEXICO
NORTH PACIFIC OCEAN
NORTH ATLANTIC OCEAN
Tropic of Cancer
Equator
Tropic of Capricorn
Antarctic Circle
GUYANA
SURINAME
FRENCH GUIANA (FR)
COLOMBIA
VENEZUELA
ECUADOR
PERU
BRAZIL
BOLIVIA
PARAGUAY
CHILE
ARGENTINA
URUGUAY
WESTERN SAMOA
TONGA
SOUTH PACIFIC OCEAN
SOUTH ATLANTIC OCEAN
THE BAHAMAS
CUBA
MEXICO
BELIZE
GUATEMALA
HONDURAS
EL SALVADOR
NICARAGUA
COSTA RICA
PANAMA
COLOMBIA
VENEZUELA
JAMAICA
HAITI
DOMINICAN REPUBLIC
PUERTO RICO
CARIBBEAN SEA
ST. KITTS AND NEVIS
ANTIGUA AND BARBUDA
DOMINICA
MARTINIQUE
ST. LUCIA
ST. VINCENT AND THE GRENADINES
BARBADOS
GRENADA
TRINIDAD AND TOBAGO
Scale: 1 to 125,000,000
0 1000 2000 Miles
0 1000 2000 3000 Kilometers

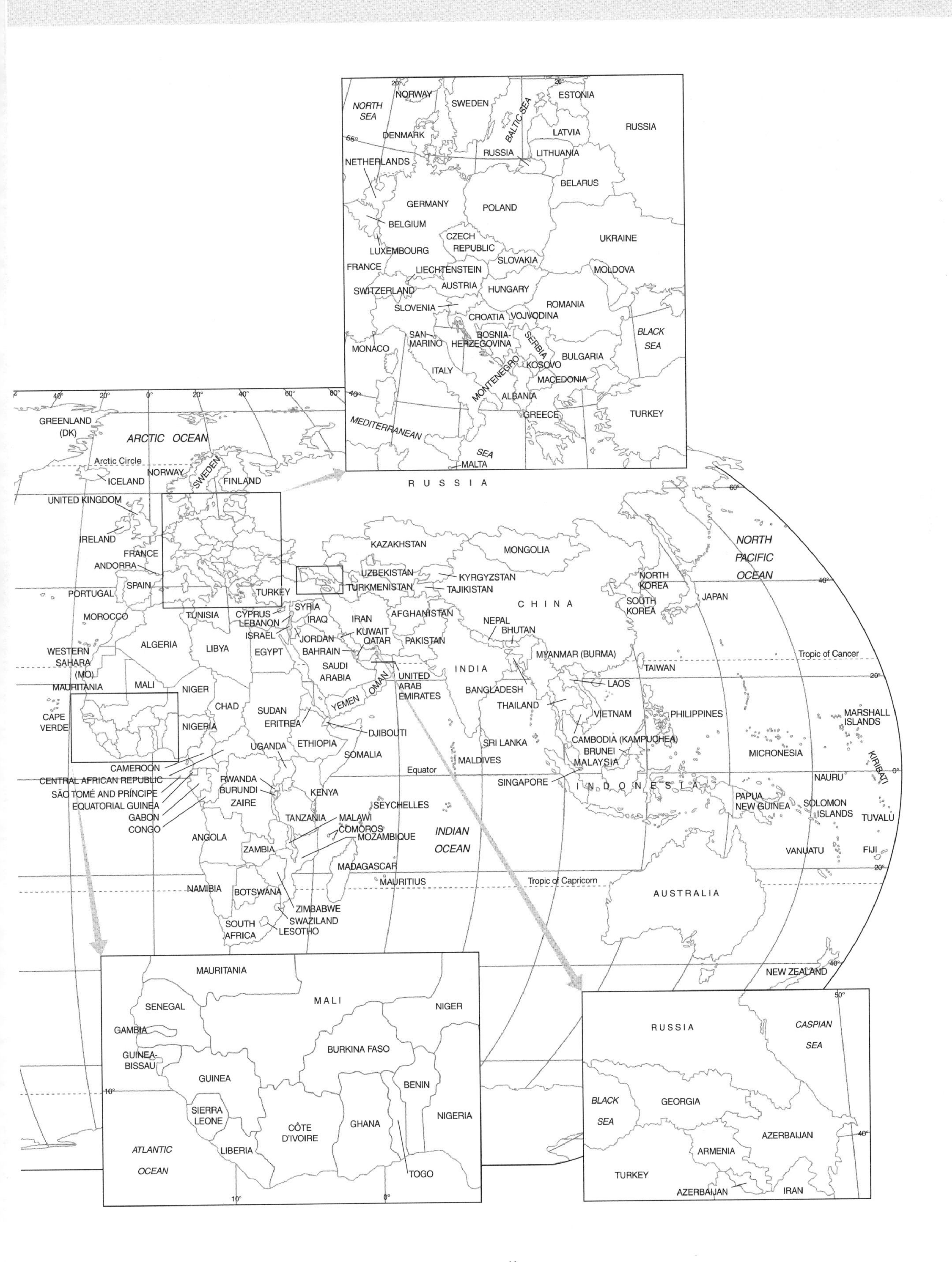

NORTH SEA
NORWAY
SWEDEN
BALTIC SEA
ESTONIA
RUSSIA
DENMARK
LATVIA
RUSSIA
LITHUANIA
NETHERLANDS
BELARUS
GERMANY
POLAND
BELGIUM
CZECH REPUBLIC
UKRAINE
LUXEMBOURG
SLOVAKIA
FRANCE
LIECHTENSTEIN
MOLDOVA
SWITZERLAND
AUSTRIA
HUNGARY
SLOVENIA
ROMANIA
CROATIA
VOJVODINA
BLACK SEA
SAN MARINO
BOSNIA-HERZEGOVINA
SERBIA
MONACO
KOSOVO
BULGARIA
ITALY
MONTENEGRO
MACEDONIA
ALBANIA
GREECE
TURKEY
MEDITERRANEAN SEA
MALTA
GREENLAND (DK)
ARCTIC OCEAN
Arctic Circle
ICELAND
NORWAY
SWEDEN
FINLAND
RUSSIA
UNITED KINGDOM
IRELAND
FRANCE
ANDORRA
PORTUGAL
SPAIN
TURKEY
KAZAKHSTAN
MONGOLIA
NORTH PACIFIC OCEAN
UZBEKISTAN
KYRGYZSTAN
TURKMENISTAN
TAJIKISTAN
NORTH KOREA
SOUTH KOREA
JAPAN
CHINA
MOROCCO
TUNISIA
CYPRUS
LEBANON
SYRIA
IRAQ
IRAN
AFGHANISTAN
NEPAL
BHUTAN
ISRAEL
JORDAN
KUWAIT
QATAR
PAKISTAN
BAHRAIN
ALGERIA
LIBYA
EGYPT
SAUDI ARABIA
INDIA
MYANMAR (BURMA)
TAIWAN
Tropic of Cancer
WESTERN SAHARA (MO)
UNITED ARAB EMIRATES
OMAN
BANGLADESH
LAOS
MAURITANIA
MALI
NIGER
YEMEN
THAILAND
VIETNAM
PHILIPPINES
MARSHALL ISLANDS
CAPE VERDE
CHAD
SUDAN
ERITREA
DJIBOUTI
NIGERIA
ETHIOPIA
SRI LANKA
CAMBODIA (KAMPUCHEA)
UGANDA
SOMALIA
BRUNEI
MICRONESIA
KIRIBATI
CAMEROON
MALDIVES
MALAYSIA
CENTRAL AFRICAN REPUBLIC
Equator
NAURU
SÃO TOMÉ AND PRÍNCIPE
RWANDA
BURUNDI
SINGAPORE
INDONESIA
EQUATORIAL GUINEA
ZAIRE
KENYA
PAPUA NEW GUINEA
SOLOMON ISLANDS
TUVALU
GABON
SEYCHELLES
CONGO
TANZANIA
MALAWI
COMOROS
MOZAMBIQUE
INDIAN OCEAN
ANGOLA
ZAMBIA
VANUATU
FIJI
MADAGASCAR
MAURITIUS
Tropic of Capricorn
NAMIBIA
BOTSWANA
AUSTRALIA
ZIMBABWE
SWAZILAND
SOUTH AFRICA
LESOTHO
NEW ZEALAND
MAURITANIA
MALI
NIGER
SENEGAL
GAMBIA
BURKINA FASO
GUINEA-BISSAU
GUINEA
BENIN
SIERRA LEONE
CÔTE D'IVOIRE
GHANA
NIGERIA
ATLANTIC OCEAN
LIBERIA
TOGO
RUSSIA
CASPIAN SEA
BLACK SEA
GEORGIA
AZERBAIJAN
ARMENIA
TURKEY
AZERBAIJAN
IRAN

UNIT 1

Geography in a Changing World

Unit Selections

1. **The Big Questions in Geography**, Susan L. Cutter, Reginald Golledge, and William L. Graf
2. **Rediscovering the Importance of Geography**, Alexander B. Murphy
3. **The Four Traditions of Geography**, William D. Pattison
4. **The Changing Landscape of Fear**, Susan L. Cutter, Douglas B. Richardson and Thomas J. Wilbanks
5. **Recreating Secure Spaces**, Ray J Dezzani and T.R. Lakshmanan
6. **Perilous Gardens, Persistent Dreams**, Rob Schultheis
7. **After Apartheid**, Judith Fein

Key Points to Consider

- Why is geography called an integrating discipline?
- How is geography related to earth science? Give some examples of these relationships.
- What are area studies? Why is the spatial concept so important in geography? What is your definition of geography?
- How has GIS changed geography and cartography?
- What does interconnectedness mean in terms of places? Give examples of how you as an individual interact with people in other places. How are you "connected" to the rest of the world?

Links: www.dushkin.com/online/
These sites are annotated in the World Wide Web pages.

Alternative Energy Institute (AEI)
http://www.altenergy.org

Mission to Planet Earth
http://www.earth.nasa.gov

Poverty Mapping
http://www.povertymap.net

Santa Fe Institute
www.santafe.edu/index.php

Solstice: Documents and Databases
http://solstice.crest.org

What is geography? This question has been asked innumerable times, but it has not elicited a universally accepted answer, even from those who are considered to be members of the geography profession. The reason lies in the very nature of geography as it has evolved through time. Geography is an extremely wide-ranging discipline, one that examines appropriate sets of events or circumstances occurring at specific places. Its goal is to answer certain basic questions.

The first question—Where is it?—establishes the location of the subject under investigation. The concept of location is very important in geography, and its meaning extends beyond the common notion of a specific address or the determination of the latitude and longitude of a place. Geographers are more concerned with the relative location of a place and how that place interacts with other places both far and near. Spatial interaction and the determination of the connections between places are important themes in geography.

Once a place is "located," in the geographer's sense of the word, the next question is, Why is it here? For example, why are people concentrated in high numbers on the North China plain, in the Ganges River Valley in India, and along the eastern seaboard in the United States? Conversely, why are there so few people in the Amazon basin and the Central Siberian lowlands? Generally, the geographer wants to find out why particular distribution patterns occur and why these patterns change over time.

The element of time is another extremely important ingredient in the geographical mix. Geography is most concerned with the activities of human beings, and human beings bring about change. As changes occur, new adjustments and modifications are made in the distribution patterns previously established. Patterns change, for instance, as new technology brings about new forms of communication and transportation and as once-desirable locations decline in favor of new ones. For example, people migrate from once-productive regions such as the Sahel when a disaster such as drought visits the land. Geography, then, is greatly concerned with discovering the underlying processes that can explain the transformation of distribution patterns and interaction forms over time. Geography itself is dynamic, adjusting as a discipline to handle new situations in a changing world.

Geography is truly an integrating discipline. The geographer assembles evidence from many sources in order to explain a particular pattern or ongoing process of change. Some of this evidence may even be in the form of concepts or theories borrowed from other disciplines. The first article in this book raises what are considered to be the big questions in geography, and the second stresses the importance of geography as a discipline and proclaims its "rediscovery."

Throughout its history, four main themes have been the focus of research work in geography. These themes or traditions, according to William Pattison in "The Four Traditions of Geography," link geography with earth science, establish it as a field that studies land-human relationships, engage it in area studies, and give it a spatial focus. Although Pattison's article first appeared over 30 years ago, it is still referred to and cited frequently today. Much of the geographical research and analysis engaged in today would fall within one or more of Pattison's traditional areas, but new areas are also opening for geographers.

The next two selections were chosen from the highly acclaimed book, *The Geographical Dimensions of Terrorism.* The first, "The Changing Landscape of Fear," states that geography can play a key role in both understanding and combating terrorism. The second, "Recreating Secure Spaces," deals with the vulnerability of transportation systems in the era of global terrorism. In "Perilous Gardens, Persistent Dreams," the restoration of Afghanistan following years of war and drought is discussed.

"After Apartheid" recounts the positive changes that have occurred in South Africa following the end of the repressive regime.

The Big Questions in Geography

In noting his fondness for geography, John Noble Wilford, science correspondent for *The New York Times*, nevertheless challenged the discipline to articulate those big questions in our field, ones that would generate public interest, media attention, and the respect of policymakers. This article presents our collective judgments on those significant issues that warrant disciplinary research. We phrase these as a series of ten questions in the hopes of stimulating a dialogue and collective research agenda for the future and the next generation of geographic professionals.

Susan L. Cutter

University of South Carolina

Reginald Golledge

University of California, Santa Barbara

William L. Graf

University of South Carolina

Introduction

At the 2001 national meeting of the Association of American Geographers (AAG) in New York City, the opening session featured an address by John Noble Wilford, science correspondent for *The New York Times*. In very candid language, Wilford challenged the discipline to articulate the big questions in our field—questions that would capture the attention of the public, the media, and policymakers (Abler 2001). The major questions posed by Wilford's remarks include the following: Are geographers missing big questions in their research? Why is the research by geographers on big issues not being reported? And what role can the AAG play in improving geographic contributions to address big issues?

First, geographers are doing research on some major issues facing modern society, but not all of them. Geographic thinking is a primary component of the investigation of global warming, for example. Products of that research seen by decision makers and the public often take the form of maps and remote sensing images that explain the geographic outcomes of climate change. Geographic approaches are at the heart of much of the analysis addressing natural and technological hazards, with public interaction taking place through the mapping media. Earthquake, volcanic, coastal, and riverine hazards are all subject to spatial analysis that has become familiar to the public. The terrorist attacks of 11 September 2001 have stimulated new interest in geographic information systems that can be used in response to hazardous events and as guidance in emergency preparedness and response (Figure 1).

In addition to these recent challenges, however, there are major issues that geographers are not addressing adequately at the present time, as illustrated by the accounting that follows in this article. A primary reason for the disconnect between capability to help solve problems and the application of those skills for many major issues is the sociology of the discipline of geography. The majority of AAG members, for example, are academicians, and their agendas and reward structures are targeted at specialized research deeply buried in paradigms that are obscure to decision makers and the public. Additionally, this social structure tends to lead geographic researchers into investigations on small problems that can be solved quickly, produce professional publications, and support a drive for promotion and tenure, rather than investigating more complex, bigger problems that are not easily or quickly solved and do not necessarily lead to academic publications of a type the genre usually demands.

With few exceptions, those geographers outside the university setting are scattered and work individually, in small groups, or as members of larger interdisciplinary teams for governmental agencies, businesses, or private organizations. Because there are few true "institutes" of geographic research, it is difficult to focus geographic energy on big problems. Many geographers in these settings are responding to immediate and short-term demands on their time and talents, rather than leading the larger-scale investigations.

The work in which many geographers engage to address major problems is not reported for two reasons: it does not fit the classic mold for the research journals where geographers get their greatest career awards, and work related to policy often emerges without attribution to the researchers of origin. A significant example that illustrates this point is the work of the Committee on Geography of the Board on Earth Sciences and Resources in the National Research Council (NRC). The committee oversees study committees, which produce geographic studies and reports to guide the federal government in a wide variety of issues that qualify as big questions. Recent work, contributed to primarily by geographers and accomplished from a geographic perspective, includes advice to the U.S. Geological Survey on

Figure 1 *Manhattan, New York, before the terrorist attacks of 11 September 2001 (left), and after (right). Photos by S. L. Cutter.*

reformulating its research programs to address geographic issues entangled in urban expansion, hazards, and mapping. In other cases, one study committee is producing direction for the federal government on what decision makers and the general public need to know about the world of Islam, while another is investigating transportation issues related to urban congestion and the development of livability indicators. Other geographers participate in the Water Science and Technology Board of the NRC, with recent contributions including the use of the watershed concepts in ecosystem management and the role of dams in the security of public water supplies. Another example involves a geographer-led multidisciplinary group to investigate spatial thinking, and another geographer led a major effort in global mapping. In all of these cases, geographers play a central role, but the product of their work is ascribed only to an organization (the NRC), and individuals are recognized only in lists of contributors. If the reports successfully influence policy, the decision makers who actuate that policy take credit for the process, rather than the original investigators who made the recommendations.

The AAG plays a role in stimulating the research that addresses big questions of importance to modern society by recognizing such work and publicizing it. It may be that individual researchers will be more willing to undertake such research if their work is recognized by their colleagues in the discipline as being important and worthy of praise. The AAG can influence the National Science Foundation, the National Endowment for the Humanities, the National Institutes of Health, the National Geographic Society, and other funding sources to channel attention and resources to individuals or teams examining the big questions. Individual geographers are not likely to be able to exert much influence, except when they serve on review panels for these organizations, but the AAG can exert its influence from its steady and visible presence in Washington.

In trying to identify those issues that might qualify as big questions (Table 1), we have included wide-ranging concepts that encompass some conceptual issues (such as scale), but also point out specific topical areas that seem to demand particular attention at the moment. We argue that such diverse big ideas belong in this accounting because, in the end, they are related to each other and mutually supportive. Some of these big questions may be obscure to the public, but most of them are familiar to researchers and policymakers alike, who have already begun to address them. There is little hope that any collection of big questions can identify problems of equal "bigness," but the ones we have identified all seem to warrant teams of researchers and significant funding rather than following the discipline's usual mode of a single or small group of investigators with funding limited to one or two years in duration. The communication of geographic research findings to the public in thoughtful, useful ways represents a major challenge. This challenge, by itself, can also be regarded as yet another big problem facing the discipline. With these introductory comments in mind, we now turn to those questions that we feel are important for the geographic community to address.

What Makes Places and Landscapes Different from One Another and Why Is This Important?

This first question goes to the core of the discipline the relevance of similarities and differences among people, places, and regions. What is the nature of uneven economic development and what can geography contribute to understanding this phenomenon? More specifically, how can national and global policies be implemented in a world that is increasingly fragmented politically, socially, culturally, and environmentally?

To elaborate on this question, we accept an assumption that the human mind is not constructed to handle large-scale continuous

Table 1 Big Questions in Geography

1. What makes places and landscapes different from one another, and why is this important?
2. Is there a deeply held human need to organize space by creating arbitrary borders, boundaries, and districts?
3. How do we delineate space?
4. Why do people, resources, and ideas move?
5. How has the earth been transformed by human action?
6. What role will virtual systems play in learning about the world?
7. How do we measure the unmeasurable?
8. What role has geographical skill played in the evolution of human civilization, and what role can it play in predicting the future?
9. How and why do sustainability and vulnerability change from place to place and over time?
10. What is the nature of spatial thinking, reasoning, and abilities?

chaos. Nor does it function optimally when dealing with large-scale perfect uniformity. Between these two extremes there is variability, which is the dominant characteristic of both the natural world and the human world. To understand the nature of physical and human existence, we need to examine the occurrence and distribution of variability in various domains. For geographers, this examination has involved exploring the nature of spatial distributions, patterns, and associations, examining the effects of scale, and developing modes of representation that best communicate the outcomes of these explorations. In the course of this search for understanding of the essentials of spatial variation, geographers have attempted to comprehend the interaction between physical and human environments, how people adapt to different environments, and how knowledge about human-environment relations can be communicated through appropriate representational media.

Even in the absence of humans, the earth and the phenomena found on this planet are incredibly diverse. Variability is widespread; uniformity is geographically restricted. Determining the nature and occurrence of variability and uniformity are at the heart of the discipline of geography. No other area of inquiry has, as its primary goal, discovering, representing, and explaining the nature of spatial variability in natural and human environments at scales beyond the microscopic and the figural (body space) such as vista, environmental, or gigantic and beyond (Montello 1993). Most geography has been focused on vista, environmental or gigantic scales, but some (e.g., cognitive behavioral) emphasizes figural scale. Finding patterns or trends towards regularity at some definable scale amidst this variability provides the means for generalizing, modeling, and transferring knowledge from one spatial domain to another. Law-like and theoretical statements can be made, and confidence in the relevance of decisions and policies designed to cope with existence can be determined.

Among other things, geographers have repeatedly found, at some scales, spatial regularity in distributions of occurrences that seem random or indeed chaotic at other scales. Sometimes this results from selecting an appropriate scale and format for summarizing and representing information. Examples include using very detailed environmental-scale data to discover the topologic properties of stream networks, or establishing the regular and random components of human settlement patterns in different environments.

Realizing the spatial variability in all phenomena is a part of the naïve understanding of the world. Being able to explain the nature of variability is the academic challenge that drives the discipline of geography. Like other scientists, geographers examine variability in their search for knowledge and understanding of the world we live in, particularly in the human environment relations and interactions that are a necessity for our continued existence.

Is There a Deeply Held Human Need to Organize Space by Creating Arbitrary Borders, Boundaries, and Districts?

Humans, by their very nature, are territorial. As human civilizations grew from hunter gatherers to more sedentary occupations, physical manifestation of the demarcation of space ensued. Hadrian's Wall kept the Scots and Picts out, the Great Wall of China protected the Ming Dynasties from the Mongols, and the early walled cities of Europe protected those places from barbarians and other acquisitive sociocultural groups.

At a more limited scale, internal spaces in cities were also divided, often based on occupation and/or class. As civilizations grew, space was organized and reorganized into districts that supported certain economic activities. City-states begat nation-states, and eventually most of the world was carved up into political spaces. Nation-states required borders and boundaries (all involving geography), as land and the oceans (and the resources contained within) were carved up into non-equal units. Within nations, land partitioning has been a factor in the decline of environmental quality. For example, the erection of barbed-wire fencing on the Great Plains to separate farming and ranching homesteads from each other did more to hasten the decline of indigenous species and landscape degradation than any other invention at the time (Worster 1979, 1993).

The modern equivalent of the human need and desire for delineating space is the notion of private property. Suburban homes with tall fences between neighbors, for example, help foster the ideal of separation from neighbors and disengagement from the community, both predicated on the need to protect "what's mine" (and of course the ubiquitous property value) as well as providing a basic need for privacy. The tendency for the rich to get richer and the poor to get poorer also applies to the values of these divided properties. The diffusion of democratically controlled, market based economies to much of the globe increases the significance of research that explores why we divide space. Pressing research questions include, for example: Are ghettos bursting with poverty-level inhabitants an inevitable consequence of democratic capitalist societies? Are such societies amenable to concerns for social justice? And how would such concerns influence the patterns and distributions of living activities?

We also lack some of the basic understanding of how the physical delineation of space affects our perception of it. Furthermore, we need better knowledge of how perceptions of physical space alter social, physical, and environmental processes. Finally, has globalization changed our view of the social construction of space? Does physical space still support spatial relations and spatial interactions, or are they becoming somewhat independent, as may be the case in social space, intellectual space, and cyberspace? How will the interactions between people, places, and regions change as our view of space (and time, for that matter) changes?

How Do We Delineate Space?

Once we understand *why* we partition space, we face a closely related issue: *how* do we do it? The definition of regions by drawing boundaries is deceptively simple. The criteria by which we delineate space have far-reaching consequences, because the resulting divisions of space play a large role in determining how we perceive the world. A map of the United States showing the borders of the states, for example, evokes a very different perception of the nation than a similar scale map showing the borders of the major river basins. A further difference in perception is created if the map shows major rivers as networks rather than as basins, and the resulting difference between perception of networks and perception of regions can direct knowledge and its application in divergent ways (National Research Council 1999, x). For example, should we conduct pollution oriented research on rivers or on watersheds, or on the state administrative units that potentially might control pollution? What are the implications of our choice of geographic framework?

The logical, rational delineation of spaces on the globe depends on the criteria to be used, but geographic research offers few established, widely accepted rules about what these criteria should be or how they might be employed. The designation of political boundaries without respect to ethnic cultures has wrought havoc in much of post colonial Africa and central Europe, for example, but geographers have not yet offered workable alternatives that account for the complexities of multicultural populations. In natural-science research and management, a major issue is the establishment of meaningful regions that can be aggregated together to scale up, or that can be disaggregated to scale down. Natural scientists also experience significant difficulty in designing compatible regions across topical subjects. For example, the blending of watersheds, ecosystems, and ranges of particular species poses significant problems in environmental management. Adding to the complexity from a management and policy perspective is the tangle of administrative regions, whose boundaries are often derived from political boundaries rather than natural ones. Recognition of these problems is easy but offering thoughtful geographic solutions to them is not.

Geographers have much to contribute to the delineation of space by developing new knowledge and techniques for defining subdivisions of earth space based on specific criteria, including economic efficiency, compatibility across applications, ease of aggregation and disaggregation, repeatability, and universality of application. Geographers need to develop methods for delineating space that either resist change over time or accommodate temporal changes smoothly.

A continuing example of delineating space that has important political implications is the process of defining American congressional districts once each decade based on the population census. The need for fair representation, relative uniformity in population numbers in each district, recognition of traditional communities, and accommodation of changing population distributions comprise some of the criteria that need not equate to partisan politics in constructing at least the first approximation of redrawn district boundaries (Monmonier 2001). Some states have nonpartisan commissions to delineate the districts, yet geography provides very little substantive advice on the subject to guide such groups.

Why do People, Resources, and Ideas Move?

One of the fundamental concepts in geography is the understanding that goods, services, people, energy, materials, money, and even ideas flow through networks and across space from place to place. Although geography faces questions about all these movements, one of the most pressing questions concerns the movement of people. We have some knowledge about the behavior of people who move their residences from one place to another, and we can observe obvious economic forces leading to the migration of people toward locations of relative economic prosperity. However, we have much less understanding about the episodic movements of people in cities. In most developed countries, the congestion of vehicular traffic has become a significant negative feature in assessing the quality of life, and in lesser-developed countries the increasing number of vehicles used in the context of inadequate road networks results in frustrating delays. Geography can and should address fundamental issues such as the environmental consequences of the decision to undertake laborious journeys to work (e.g., contributions of vehicle exhaust to air pollution, the possible environmental changes induced by telecommuting, and the need for alternative-fuel, low-pollution vehicles). In addition to understanding the environmental consequences of daily moves, the discipline has much to offer in describing, explaining, and predicting the sociocultural consequences of these decisions.

The flow of vehicles on roadways involves obvious physical networks, but there are other flows demanding attention that operate through more abstract spaces. The diffusion of culture—particularly "Western" culture, with its emphasis on materialism and individualism—is one of the leading edges of globalization of the world economy. Geographers must begin to address how these social, cultural, and economic forces operate together to diffuse, from a few limited sources, an extensive array of ideas and attitudes that are accepted by a diverse set of receiving populations. Even if such diffusion takes place through digital space, it probably does so in a distinctive geography that we should understand if we are to explain and predict the world in the twenty-first century.

The electrical energy crisis of 2001 made us aware, quite vividly, of the finiteness of nonrenewable resources such as oil and gas and of the difficulties in their distributions. We have already consumed more than 50 percent of the world's known reserves of these resources. Historically, as one energy source has replaced another (as when coal power replaced water power), there have been changes in the locational patterns, growth, importance of settlements, and significance of regions. Examples include the decline of heavy industrial areas into "rust belts" and their replacement with service- and information-based centers that have more locational flexibility. As current energy sources change, what will happen to urban location and growth? Will the geopolitical power structure of the world change markedly? For example, will the countries that are part of the Organization of the Petroleum Exporting Countries (OPEC) retain their global economic power and political strength? Will existing populations and settlements decline, or relocate to alternative sources of energy? What will be the geographic configuration of the economic and political power that goes with such changes?

Finally, the more physically oriented flows, such as those of energy and materials, present a demanding set of questions for geographers. While geochemists are deriving the magnitudes of elemental fluxes of such substances as carbon and nitrogen, for example, it is incumbent on geographers to point out that these fluxes do not take place in aspatial abstract ways, but rather in a physically and socially defined landscape that has important locational characteristics. In other words, although there may very well be an understanding of the amounts of nitrogen circulating from earth to oceans to atmosphere, that circulation is not everywhere equal. How does human management affect the nitrogen and other elemental cycles? What explains its geographic vari-

Figure 2 *A local example of transformations brought about in the natural world by humans. The lower Sandy River of western Oregon appears to be a pristine river, but it has radically altered water, sediment, and biological systems because of upstream dams. Photo by W. L. Graf.*

ability? How does that variability change in response to controls not related to human intervention? This leads to our next big question.

How Has the Earth Been Transformed by Human Action?

Humans have altered the earth, its atmosphere, and its water on scales ranging from local to global (Thomas 1956; Turner et al. 1990). At the local scale, many cities and agricultural landscapes represent nearly complete artificiality in a drive to create comfortable places in which to live and work, and to maximize agricultural production for human benefit. The transformations have also had negative effects at local scales, such as altering the chemical characteristics of air and water, converting them into media that are toxic for humans as well as other species. At regional scales, human activities have resulted in wholesale changes in ecosystems, such as the deforestation of northwest Europe over the past several centuries, a process that seems to be being replicated in many tropical regions today. At a global scale, the introduction of industrial gases into the atmosphere plays a still emerging role in global climate change. Taken together, these transformations have had a geographically variable effect that geographers must better define and explain. Dilsaver and Colten (1992, 9) succinctly outlined the basic questions almost a decade ago: How have human pursuits transformed the environment, and how have human social organizations exerted their control over environments? Graf (2001) recently asked how we can undo some of our previous efforts at environmental change and control.

In many instances, this explanation of variation might emphasize the physical aspects of changes, or understanding the underlying dynamics of why the changes occur (Dilsaver, Wyckoff, and Preston 2000). Wide-ranging assessments of river basins, for example, must rely on a plethora of controlling factors ranging from land use to water, sediment, and contaminant movements. Geographers must employ more complicated and insightful approaches, however, to truly understand why transformations vary from place to place, largely in response to the connection between the biophysical environment and the human society that occupies it. Understanding this delicate interplay between nature and society as a two-way connection can lead us to new knowledge about social and environmental landscapes, but it can also help us make better decisions on how to achieve future landscapes that are more often transformed in nondestructive ways.

One of the primary issues facing many societies in their relationship with their supporting environments is how much of the biophysical world should be left unchanged, or at least changed to the minimal degree possible. The amount of remaining "natural" landscape in many nations is small—probably less than 5 percent of the total surface—so time is growing short to decide what areas should be set aside and preserved (Figure 2). Not only do these preservation decisions affect land and water surfaces; they also profoundly affect nonhuman species that use the surfaces for habitat. If human experience is enriched by diverse ecosystems, then the decline in biodiversity impoverishes humanity as well. Which areas should be preserved and why? How should preserved areas be linked with one other? How can public and private property productively coexist with nearby preserved areas?

What Role Will Virtual Systems Play in Learning about the World?

Stated another way, what will virtual systems allow us to do in the future that we cannot do now? What new problems can be pursued (Golledge forthcoming)? Providing an answer opens a Pandora's box of questions concerning the geographic impacts of new technologies (Goodchild 2000). What new multimodal interfaces for interpreting visualized onscreen data need to be developed in order to overcome current technological constraints of geographic data visualization? Can we produce a virtual geography? Do we really want to?

One serious problem that deserves immediate attention is the examination of the geographic implications of the development of economies and societies based on information technology. In particular, the sociospatial implications of an increasing division between the digital haves and have-nots demand attention. Pursuing such a problem will require answering questions about the geographic consequences of employment in cyberspace and its implication for human movements such as migration, intraurban mobility, commuting, and activity-space restructuring. The current extensive demand for and use of transportation for business purposes may need to be re-examined. It may be argued that, in the world of business communication, geographic distance is a decreasingly important factor, because both digital and visual interaction can take place at the click of a mouse button without the need for person-to-person confrontation. If this is so, what are the longer-term impacts for living and lifestyles, and how could the inhabitation and use of geographic environments be affected? If this is true, why is it that we see dramatic concentrations of cyber-businesses in a few areas, similar to the locational behavior of pre-digital industries? Are Silicon Valley in California and Route 128 in Massachusetts simply the "rust belts" of the future?

Research has shown that the most effective way of learning about an environment is by directly experiencing it, so that all sensory modalities are activated during that experience (Figure 3) (Gale 1984; Lloyd and Heivly 1987; MacEachren 1992). However, many places are distant or inaccessible to most people. The interior of the Amazon rainforest, the arctic tundra of northern Siberia, Himalayan peaks, the interior of the Sahara desert, Antarctica and the South Pole, the barrios of Rio de Janeiro, and the Bosnian highlands can become much closer to us. Satellite imagery provides detailed digitized imagery of these

Figure 3 *Exploring immersive virtual worlds with equipment developed between 1992 and 2001, showing the original and the miniaturized versions of a GPS-driven auditory virtual environment at the University of California, Santa Barbara. The more cumbersome 1992 version is shown on the left, with the reduced 2001 version on the right. Psychologist Jack Loomis and associates developed the system, demonstrated here by author Reginald Golledge. Photo courtesy of R. Golledge.*

places. A problem awaiting solution is how to use this extensive digital database to build virtual systems that will allow immersive experiences with such environments. Problems of motion-sickness experienced by some people in immersive systems need to be solved; assuming this will be achieved, virtual reality could become the laboratory of the future for experiencing different places and regions around the world.

Discovering how best to deal with problematic futures, on earth or on other planets, is definitely one of the big problems facing current and future geographers. Many land use planning, transportation, and social policies are made on an "if ______ then ______" basis. Because we are unable to change the world experimentally, we need to investigate other ways of observing environmental events and changes. Examples include changing a street for vehicles to a pedestrian mall to explore human movement behavior, or experiencing the action and consequences of snow or mud avalanches in tourist-dependent alpine environments. What more can we learn by building and manipulating virtual environments? In a virtual system, we can raise local pollution levels, accelerate global warming, change sea levels by melting ice caps, or simulate the impacts of strictly enforcing land conversion policies at the rural-urban fringes of large cities. In the face of an increasingly international economy and globalization of environmental issues, there is a need to develop a way to explore possible scenarios before implementing policies theoretically designed to deal with global (or more local) problems.

How Do We Measure the Unmeasurable?

Geography is normally practiced at local to national scales at which we can get a clear sense of the existence or development of patterns and processes. People, landscapes, and resources are not evenly distributed on the earth's surface, so we begin with a palette that is diverse. How can we accommodate such diversity in policies to avoid winners and losers? Economists, for example, assume away all spatial variability in their economic models. What happens to general models when space is introduced? How can we transform from the local to the global and vice versa? The question of scale transformation, especially the calibration of large-scale global circulation models or the development of climate-impact models globally with local or regional applications is a major area in which geography can contribute and is playing a leading role (AAG GCLP Group forthcoming).

We need to develop more compatible databases that have an explicit geographic component, with geocoded data that permit us to scale up and scale down as the need arises. Data collection, archiving, and dissemination all are areas that require our expertise, be it demographic data, environmental data, or land-use data. The large question is, how do we maintain a global information system that goes beyond the petty tyrannies of nation-states (and the need to protect information for "security" reasons), yet protects individuals' right to privacy? The selective use of remote-sensing techniques to monitor environmental conditions has been helpful in understanding the linkages between local activities and global impacts. However, can we use advanced technology to support demographic data collection and analyses and still maintain safeguards on privacy protection (Liverman et al. 1998)? For example, recent Supreme Court decisions have placed important legal protections on the use of thermal infrared sensors in public safety.

Another series of issues involves the aggregation of human behaviors. How can we geographically aggregate data along a set of common dimensions to insure its representation of reality and get around the thorny issues of averaging and the mean-areal-center or modified areal-unit problems? We often use techniques to handle aggregated populations and areas that in fact, depart from reality, creating a type of artificial environment. Unfortunately, public policies all too often are based on these constructed realities, thus further exacerbating the distribution of goods, services, and resources. What new spatial statistical tools do we need to address this concern?

Lastly, in a post-11-September world, how do we measure the geography of fear? Does the restriction of geographic data (presumably for national security reasons) attenuate or amplify fear of the unknown? The discipline requires the open access to information and data about the world and the people who live there. Data access will be one of the key issues for our community to address in the coming years.

What Role Has Geographical Skill Played in the Evolution of Human Civilization, and What Role Can It Play in Predicting the Future?

Is there a necessary geographic base to human history? If so, how can we improve our ability to predict spatial events and events that have spatial consequences that will fundamentally shape the future? Can we develop the geographic equivalent of leading economic indicators?

From the early cradles of civilization in Africa and Asia, humankind gradually colonized the earth. This process of redistributing people in space (migration) was caused by population growth, resource exhaustion, attractive untapped resources, environmental change, environmental hazard, disease, or invasion and succession by other human groups. But what skills and abilities were required to ensure success in relocation movements? Were the movements random or consciously directed? If they were directed, then what skills and/or abilities were required by explorers, leaders, and followers to ensure success? What criteria had to be satisfied before resettlement was possible? What new geographic skills and abilities have been developed throughout human history, and which ones have deteriorated or disappeared? Have geographic skills and abilities been maintained equally in males and females? If not, what developments in the evolution of human civilizations have mediated such losses or changes?

While we know much about human history, we know little about the geographical basis of world history, and we know little of the extent to which the presence or absence of geographic knowledge played a significant part in historical development. For example, would Napoleon's invasion of Russia been more successful had skilled and knowledgeable geographers counseled him on the route chosen and the appropriate season for movement? Historians often tell us that understanding the past is the key to knowing the present and to successfully predicting the future. We cannot fully understand the past if we ignore or diminish the importance of environmental diversity and knowledge about those variations that are the result of spatial and geographic thinking and reasoning. A similar argument can be made for predicting future events and behaviors. What geographic knowledge is likely to be important in prediction? Must we rely on assumptions about uniform environments, population characteristics, tastes and preferences, customs, beliefs, and values? Such a procedure is precarious at best. However, we do not currently know how to incorporate geographic variability into our models, or indeed what variables should be incorporated into predictive models. Achieving such a goal is a necessary part of increasing our very limited predictive capabilities.

How and Why Do Sustainability and Vulnerability Change from Place to Place and over Time?

Historically, geography was an integrative science with a particular focus on regions. It then switched from breadth to depth, with improvements in theory, methods, and techniques. We are now returning to that earlier perspective as we look for common ground in the interactions between human systems and physical systems. Increasing population pressures, the regional depletion or total exhaustion of resources, environmental degradation, and rampant development are processes that affect the sustainability of natural systems and constructed environments. There is a movement toward the integration of many different social and natural science perspectives into a field called sustainability science (Kates et al. 2001). Understanding what constrains and enhances sustainable environments will be an important research theme in the future. How can we maintain and improve the quality of urban environments for general living (social, economic, and environmental conditions)? How long can the processes of urban and suburban growth continue without deleterious and fundamental changes in the landscape and the escalation in costs of environmental restoration? Suburban sprawl is already a major policy issue. What is the long-term impact on human survival of the constant usurpation of agricultural land by the built environment? How long can we continue slash and burn agriculture in many parts of the tropical world? What triggers the environmental insecurity of nations, and how does this lead to armed conflicts and mass migrations of people? How have these processes varied in time and space? What are the greatest threats to the sustainability of human settlements, agriculture, energy use, for example and how can we mitigate or reduce those threats (NRC 2000)?

Nonsustainable environments enhance the effect of risks and hazards and ultimately increase both biophysical and social vulnerability, often resulting in disasters of one kind or another. When societies or ecosystems lack the ability to stop decay or decline and they do not have the adequate means to defend against such changes, there can be potentially catastrophic results. Examples include the environmental degradation of the Aral Sea, the increasing AIDS pandemic, and the human and environmental costs of coastal living (Heinz Center 2000). Vulnerability can be thought of as a continuum of processes, ranging from the initial susceptibility to harm to resilience (the ability to recover) to longer-term adaptations in response to large-scale environmental changes (Cutter, Mitchell, and Scott 2000). These processes manifest themselves at different geographic scales, ranging from the local to the global. What is the threshold when vulnerability ceases to become something we can deal with and becomes something we cannot? At what point does the built environment or ecosystem extend beyond its own ability to recover from natural or social forces?

What Is the Nature of Spatial Thinking, Reasoning, and Abilities?

Geographic knowledge is the product of spatial thinking and reasoning (Golledge 2002b). These processes require the ability to comprehend scale changes; transformations of phenomena, or representations among one, two, and three spatial dimensions. They also require understanding of: the effect of distance, direction, and orientation on developing spatial knowledge; the nature of reference frames for identifying locations, distributions and patterns; the nature of spatial hierarchies; the nature of forms by extrapolating from cross sections; the significance of adjacency and nearest neighbor concepts; the spatial properties of density, distance, and density decay; and the configurations of patterns and shapes in various dimensions and with differing degrees of completeness. It also requires knowing the implications of spatial association and understanding other concepts not yet adequately articulated or understood. What geography currently lacks is an elaboration of the fundamental geographic concepts and skills that are necessary for the production and communication of spatial and geographic information. In the long run, this will be needed before geography can develop a well-articulated knowledge base of a type similar to other human and physical sciences.

Conclusion

In the American Declaration of Independence, Thomas Jefferson wrote that among the most basic of human rights are life, liberty, and the pursuit of happiness. Each of these rights is played out upon a geographic stage, has geographic properties, and operates as a geographical process. Geography, as a field of knowledge and as a perspective on the world, has paid too little attention to these grand ideas, and they are fertile ground for the seeds of new geographic research. How and why does the opportunity for the pursuit of happiness vary from one place to another, and does the very nature of that pursuit change geographically?

In pursuit of answers to the big questions articulated above, we will inevitably need to think about doing research on problems such as:

- What are the spatial constraints on pursuing goals of life, liberty, and the pursuit of happiness?
- What are our future resource needs, and where will we find the new resources that have not, at this stage, been adequately explored?
- When does geography start and finish? Does it matter?
- What are likely to be the major problems in doing the geography of other planets?
- Will cities of the future remain bound to the land surface, or will they move to what we now consider unlikely or exotic locations (under water or floating in space)?

The big questions posed here are not all encompassing. They represent our collective judgments (and biases) on what issues are significant for the discipline, and those that should provide a focus for our considerable intellectual capital. Not everyone will agree with us, nor should they. We view this article as the beginning of a dialogue within the discipline as to what are the probable big questions for the next generation of geographers.

Literature Cited

Abler, R. F. 2001. From the meridian—Wilford's "science writer's view of geography." *AAG Newsletter* 36 (4): 2, 9.

Association of American Geographers (AAG) Global Change in Local Places (GCLP) Research Group. Forthcoming. *Global change and local places: Estimating, understanding, and reducing greenhouse gases*. Cambridge, U.K.: Cambridge University Press.

Cutter, S. L., J. T. Mitchell, and M. S. Scott. 2000. Revealing the vulnerability of people and places: A case study of Georgetown County, South Carolina. *Annals of the Association of American Geographers* 90:713–37.

Dilsaver, L. M., and C. E. Colten, eds. 1992. *The American environment: Interpretations of past geographies*. Lanham, MD: Rowan and Littlefield Publishers.

Dilsaver, L. M., W. Wyckoff, and W. L. Preston. 2000. Fifteen events that have shaped California's human landscape. *The California Geographer* XL: 1–76.

Gale, N. D. 1984. Route learning by children in real and simulated environments. Ph.D. diss., Department of Geography, University of California, Santa Barbara.

Golledge, R. G. 2002. The nature of geographic knowledge. *Annals of the Association of American Geographers* 92 (1): 1–14.

—. Forthcoming. *Spatial cognition and converging technologies*. Paper presented at the Workshop on Converging Technology (NBIC) for Improving Human Performance, sponsored by the National Science Foundation. Washington, D.C. In press.

Goodchild, M. F. 2000. Communicating geographic information in a digital age. *Annals of the Association of American Geographers* 90:344–55.

Graf, W. L. 2001. Dam age control: Restoring the physical integrity of America's rivers. *Annals of the Association of American Geographers* 91:1–27.

Heinz Center. 2000. *The hidden costs of coastal erosion*. Washington, D.C.: The H. John Heinz III Center for Science, Economics and the Environment.

Kates, R. W., W. C. Clark, R. Corell, J. M. Hall, C. C. Jaeger, I. Lowe, J. J. McCarthy, H. J. Schnellnhuber, B. Bolin, N. M. Dickson, S. Faucheux, G. C. Gallopin, A. Grubler, B. Huntley, J. Jager, N. S. Jodha, R. E. Kasperson, A. Mabogunje, P. Matson, H. Mooney, B. Moore III, T. O'Riodan, and U. Svedin. 2001. Sustainability science. *Science* 292:641–42.

Liverman, D., E. F. Moran, R. R. Rindfuss, and P. C. Stern, eds. 1998 . *People and pixels: Linking remote sensing and social science*. Washington, D.C.: National Academy Press.

Lloyd, R. E., and C. Heivly. 1987. Systematic distortion in urban cognitive maps. *Annals of the Association of American Geographers* 77:191–207.

MacEachren, A. M. 1992. Application of environmental learning theory to spatial knowledge acquisition from maps. *Annals of the Association of American Geographers* 82 (2): 245–74.

Monmonier, M. S. 2001. *Bushmanders and Bullwinkles: How politicians manipulate electronic maps and census data to win elections*. Chicago: University of Chicago Press.

Montello, D. R. 1993. Scale and multiple psychologies of space. In *Spatial information theory: A theoretical basis for GIS. Lecture notes in computer science 716. Proceedings, European Conference, COSIT '93. Marciana Marina, Elba Island, Italy, September*, ed. A. U. Frank and I. Campari. 312–21. New York: Springer-Verlag.

National Research Council (NRC). 1999. *New strategies for America's watersheds*. Washington, D.C.: National Academy Press.

—. 2000. *Our common journey: A transition toward sustainability*. Washington, D.C.: National Academy Press.

Thomas, W. L., Jr., ed. 1956. *Man's role in changing the face of the earth*. Chicago: The University of Chicago Press.

Turner, B. L. II, W. C. Clark, R. W. Kates, J. F. Richards, J. T. Mathews, and W. Meyer, eds. 1990. *The earth as transformed by human action: Global and regional changes in the biosphere over the past 300 years*. Cambridge, U.K.: University of Cambridge Press.

Worster, D. E. 1979. *Dust bowl: The Southern plains in the 1930s*. Oxford: Oxford University Press.

—. 1993. *The wealth of nature: Environmental history and the ecological imagination*. New York: Oxford University Press.

SUSAN L. CUTTER is Carolina Distinguished Professor, Department of Geography, University of South Carolina, Columbia, SC 29208. E-mail: scutter@sc.edu. She served as president of the Association of American Geographers from 2000-2001, and is a fellow of the American Association for the Advancement of Science (AAAS). Her research interests are vulnerability science, and environmental hazards policy and management.

REGINALD GOLLEDGE is a Professor of Geography at the University of California, Santa Barbara, Santa Barbara, CA 93106. E-mail: golledge@geog.ucsb.edu and served as AAG president from 1999 to 2000. His research interests include various aspects of behavioral geography (spatial cognition, cognitive mapping, spatial thinking), the geography of disability (particularly the blind), and the development of technology (guidance systems and computer interfaces) for blind users.

WILLIAM L. GRAF is Education Foundation University Professor and Professor of Geography at the University of South Carolina, Columbia, SC 29208. E-mail: graf@sc.du. He served as AAG president from 1998–1999, and is a National Associate of the National Academy of Science. His specialties are fluvial geomorphology and policy for public land and water.

POINT OF VIEW

Rediscovering the Importance of Geography

By Alexander B. Murphy

As AMERICANS STRUGGLE to understand their place in a world characterized by instant global communications, shifting geopolitical relationships, and growing evidence of environmental change, it is not surprising that the venerable discipline of geography is experiencing a renaissance in the United States. More elementary and secondary schools now require courses in geography, and the College Board is adding the subject to its Advanced Placement program. In higher education, students are enrolling in geography courses in unprecedented numbers. Between 1985–86 and 1994–95, the number of bachelor's degrees awarded in geography increased from 3,056 to 4,295. Not coincidentally, more businesses are looking for employees with expertise in geographical analysis, to help them analyze possible new markets or environmental issues.

In light of these developments, institutions of higher education cannot afford simply to ignore geography, as some of them have, or to assume that existing programs are adequate. College administrators should recognize the academic and practical advantages of enhancing their offerings in geography, particularly if they are going to meet the demand for more and better geography instruction in primary and secondary schools. We cannot afford to know so little about the other countries and peoples with which we now interact with such frequency, or about the dramatic environmental changes unfolding around us.

From the 1960s through the 1980s, most academics in the United States considered geography a marginal discipline, although it remained a core subject in most other countries. The familiar academic divide in the United States between the physical sciences, on one hand, and the social sciences and humanities, on the other, left little room for a discipline concerned with how things are organized and relate to one another on the surface of the earth—a concern that necessarily bridges the physical and cultural spheres. Moreover, beginning in the 1960s, the U.S. social-science agenda came to be dominated by pursuit of more-scientific explanations for human phenomena, based on assumptions about global similarities in human institutions, motivations, and actions. Accordingly, regional differences often were seen as idiosyncrasies of declining significance.

Although academic administrators and scholars in other disciplines might have marginalized geography, they could not kill it, for any attempt to make sense of the world must be based on some understanding of the changing human and physical patterns that shape its evolution—be they shifting vegetation zones or expanding economic contacts across international boundaries. Hence, some U.S. colleges and universities continued to teach geography, and the discipline was often in the background of many policy issues—for example, the need to assess the risks associated with foreign investment in various parts of the world.

By the late 1980s, Americans' general ignorance of geography had become too widespread to ignore. Newspapers regularly published reports of surveys demonstrating that many Americans could not identify major countries or oceans on a map. The real problem, of course, was not the inability to answer simple questions that might be asked on *Jeopardy!*; instead, it was what that inability demonstrated about our collective understanding of the globe.

Geography's renaissance in the United States is due to the growing recognition that physical and human processes such as soil erosion and ethnic unrest are inextricably tied to their geographical context. To understand modern Iraq, it is not enough to know who is in power and how the political system functions. We also need to know something about the country's ethnic groups and their settlement patterns, the different physical environments and resources within the country, and its ties to surrounding countries and trading partners.

Those matters are sometimes addressed by practitioners of other disciplines, of course, but they are rarely central to the analysis. Instead, generalizations are often made at the level of the state, and little attention is given to spatial patterns and practices that play out on local levels or across international boundaries. Such preoccupations help to explain why many scholars were caught off guard by the explosion of ethnic unrest in Eastern Europe following the fall of the Iron Curtain.

Similarly, comprehending the dynamics of El Niño requires more than knowledge of the behavior of ocean and air currents; it is also important to understand how those currents are situated with respect to land masses and how they relate to other climatic patterns, some of which have been altered by the burning of fossil fuels and other human activities. And any attempt to understand the nature and extent of humans' impact on the environment requires consideration of the relationship between human and physical contributions to environmental change. The factories and cars in a city produce smog, but surrounding mountains may trap it, increasing air pollution significantly.

TODAY, academics in fields including history, economics, and conservation biology are turning to geographers for help with some of their concerns. Paul Krugman, a noted economist at the Massachusetts Institute of Technology, for example, has turned conventional wisdom on its head by pointing out the role of historically rooted regional inequities in how international trade is structured.

Geographers work on issues ranging from climate change to ethnic conflict to urban sprawl. What unites their work is its focus on the shifting organization and character of the earth's surface. Geographers examine changing patterns of vegetation to study global warming; they analyze where ethnic groups live in Bosnia to help understand the pros and cons of competing administrative solutions to the civil war there; they map AIDS cases in Africa to learn how to reduce the spread of the disease.

Geography is reclaiming attention because it addresses such questions in their relevant spatial and environmental contexts. A growing number of scholars in other disciplines are realizing that it is a mistake to treat all places as if they were essentially the same (think of the assumptions in most economic models), or to undertake research on the environment that does not include consideration of the relationships between human and physical processes in particular regions.

Still, the challenges to the discipline are great. Only a small number of primary- and secondary-school teachers have enough training in geography to offer students an exciting introduction to the subject. At the college level, many geography departments are small; they are absent altogether at some high-profile universities.

Perhaps the greatest challenge is to overcome the public's view of geography as a simple exercise in place-name recognition. Much of geography's power lies in the insights it sheds on the nature and meaning of the evolving spatial arrangements and landscapes that make up our world. The importance of those insights should not be underestimated at a time of changing political boundaries, accelerated human alteration of the environment, and rapidly shifting patterns of human interaction.

Alexander B. Murphy is a professor and head of the geography department at the University of Oregon, and a vice-president of the American Geographical Society.

The Four Traditions of Geography

William D. Pattison

Late Summer, 1990

To Readers of the *Journal of Geography:*

I am honored to be introducing, for a return to the pages of the *Journal* after more than 25 years, "The Four Traditions of Geography," an article which circulated widely, in this country and others, long after its initial appearance—in reprint, in xerographic copy, and in translation. A second round of life at a level of general interest even approaching that of the first may be too much to expect, but I want you to know in any event that I presented the paper in the beginning as my gift to the geographic community, not as a personal property, and that I re-offer it now in the same spirit.

In my judgment, the article continues to deserve serious attention—perhaps especially so, let me add, among persons aware of the specific problem it was intended to resolve. The background for the paper was my experience as first director of the High School Geography Project (1961–63)—not all of that experience but only the part that found me listening, during numerous conference sessions and associated interviews, to academic geographers as they responded to the project's invitation to locate "basic ideas" representative of them all. I came away with the conclusion that I had been witnessing not a search for consensus but rather a blind struggle for supremacy among honest persons of contrary intellectual commitment. In their dialogue, two or more different terms had been used, often unknowingly, with a single reference, and no less disturbingly, a single term had been used, again often unknowingly, with two or more different references. The article was my attempt to stabilize the discourse. I was proposing a basic nomenclature (with explicitly associated ideas) that would, I trusted, permit the development of mutual comprehension **and** confront all parties concerned with the pluralism inherent in geographic thought.

This intention alone could not have justified my turning to the NCGE as a forum, of course. The fact is that from the onset of my discomfiting realization I had looked forward to larger consequences of a kind consistent with NCGE goals. As finally formulated, my wish was that the article would serve "to greatly expedite the task of maintaining an alliance between professional geography and pedagogical geography and at the same time to promote communication with laymen" (see my fourth paragraph). I must tell you that I have doubts, in 1990, about the acceptability of my word choice, in saying "professional," "pedagogical," and "layman" in this context, but the message otherwise is as expressive of my hope now as it was then.

I can report to you that twice since its appearance in the *Journal*, my interpretation has received more or less official acceptance—both times, as it happens, at the expense of the earth science tradition. The first occasion was Edward Taaffe's delivery of his presidential address at the 1973 meeting of the Association of American Geographers (see *Annals AAG*, March 1974, pp. 1–16). Taaffe's working-through of aspects of an interrelation among the spatial, area studies, and man-land traditions is by far the most thoughtful and thorough of any of which I am aware. Rather than fault him for omission of the fourth tradition, I compliment him on the grace with which he set it aside in conformity to a meta-epistemology of the American university which decrees the integrity of the social sciences as a consortium in their own right. He was sacrificing such holistic claims as geography might be able to muster for a freedom to argue the case for geography as a social science.

The second occasion was the publication in 1984 of *Guidelines for Geographic Education: Elementary and Secondary Schools*, authored by a committee jointly representing the AAG and the NCGE. Thanks to a recently published letter (see *Journal of Geography*, March-April 1990, pp. 85–86), we know that, of five themes commended to teachers in this source,

> The committee lifted the human environmental interaction theme directly from Pattison. The themes of place and location are based on Pattison's spatial or geometric geography, and the theme of region comes from Pattison's area studies or regional geography.

Having thus drawn on my spatial, area studies, and man-land traditions for four of the five themes, the committee could have found the remaining theme, movement, there too—in the spatial tradition (see my sixth paragraph). However that may be, they did not avail themselves of the earth science tradition, their reasons being readily surmised. Peculiar to the elementary and secondary schools is a curriculum category framed as much by theory of citizenship as by theory of knowledge: the social studies. With admiration, I see already in the committee members' adoption of the theme idea a strategy for assimilation of their program to the established repertoire of social studies practice. I see in their exclusion of the earth science tradition an intelligent respect for social studies' purpose.

Here's to the future of education in geography: may it prosper as never before.

W. D. P., 1990

Reprinted from the Journal of Geography, 1964, pp. 211–216.

In 1905, one year after professional geography in this country achieved full social identity through the founding of the Association of American Geographers, William Morris Davis responded to a familiar suspicion that geography is simply an undisciplined "omnium-gatherum" by describing an approach that as he saw it imparts a "geographical quality" to some knowledge and accounts for the absence of the quality elsewhere.[1] Davis spoke as president of the AAG. He set an example that was followed by more than one president of that organization. An enduring official concern led the AAG to publish, in 1939 and in 1959, monographs exclusively devoted to a critical review of definitions and their implications.[2]

Every one of the well-known definitions of geography advanced since the founding of the AAG has had its measure of success. Tending to displace one another by turns, each definition has said something true of geography.[3] But from the vantage point of 1964, one can see that each one has also failed. All of them adopted in one way or another a monistic view, a singleness of preference, certain to omit if not to alienate numerous professionals who were in good conscience continuing to participate creatively in the broad geographic enterprise.

The thesis of the present paper is that the work of American geographers, although not conforming to the restrictions implied by any one of these definitions, has exhibited a broad consistency, and that this essential unity has been attributable to a small number of distinct but affiliated traditions, operant as binders in the minds of members of the profession. These traditions are all of great age and have passed into American geography as parts of a general legacy of Western thought. They are shared today by geographers of other nations.

There are four traditions whose identification provides an alternative to the competing monistic definitions that have been the geographer's lot. The resulting pluralistic basis for judgment promises, by full accommodation of what geographers do and by plain-spoken representation thereof, to greatly expedite the task of maintaining an alliance between professional geography and pedagogical geography and at the same time to promote communication with laymen. The following discussion treats the traditions in this order: (1) a spatial tradition, (2) an area studies tradition, (3) a man-land tradition and (4) an earth science tradition.

Spatial Tradition

Entrenched in Western thought is a belief in the importance of spatial analysis, of the act of separating from the happenings of experience such aspects as distance, form, direction and position. It was not until the 17th century that philosophers concentrated attention on these aspects by asking whether or not they were properties of things-in-themselves. Later, when the 18th century writings of Immanuel Kant had become generally circulated, the notion of space as a category including all of these aspects came into widespread use. However, it is evident that particular spatial questions were the subject of highly organized answering attempts long before the time of any of these cogitations. To confirm this point, one need only be reminded of the compilation of elaborate records concerning the location of things in ancient Greece. These were records of sailing distances, of coastlines and of landmarks that grew until they formed the raw material for the great *Geographia* of Claudius Ptolemy in the 2nd century A.D.

A review of American professional geography from the time of its formal organization shows that the spatial tradition of thought had made a deep penetration from the very beginning. For Davis, for Henry Gannett and for most if not all of the 44 other men of the original AAG, the determination and display of spatial aspects of reality through mapping were of undoubted importance, whether contemporary definitions of geography happened to acknowledge this fact or not. One can go further and, by probing beneath the art of mapping, recognize in the behavior of geographers of that time an active interest in the true essentials of the spatial tradition—*geometry* and *movement*. One can trace a basic favoring of movement as a subject of study from the turn-of-the-century work of Emory R. Johnson, writing as professor of transportation at the University of Pennsylvania, through the highly influential theoretical and substantive work of Edward L. Ullman during the past 20 years and thence to an article by a younger geographer on railroad freight traffic in the U.S. and Canada in the *Annals* of the AAG for September 1963.[4]

One can trace a deep attachment to geometry, or positioning-and-layout, from articles on boundaries and population densities in early 20th century volumes of the *Bulletin of the American Geographical Society*, through a controversial pronouncement by Joseph Schaefer in 1953 that granted geographical legitimacy only to studies of spatial patterns[5] and so onward to a recent *Annals* report on electronic scanning of cropland patterns in Pennsylvania.[6]

One might inquire, is discussion of the spatial tradition, after the manner of the remarks just made, likely to bring people within geography closer to an understanding of one another and people outside geography closer to an understanding of geographers? There seem to be at least two reasons for being hopeful. First, an appreciation of this tradition allows one to see a bond of fellowship uniting the elementary school teacher, who attempts the most rudimentary instruction in directions and mapping, with the contemporary research geographer, who dedicates himself to an exploration of central-place theory. One cannot only open the eyes of many teachers to the potentialities of their own instruction, through proper exposition of the spatial tradition, but one can also "hang a bell" on research quantifiers in geography, who are often thought to have wandered so far in their intellectual adventures as to have become lost from the rest. Looking outside geography, one may anticipate benefits from the readiness of countless persons to associate the name "geography" with maps. Latent within this readiness is a willingness to recognize as geography, too, what maps are about—and that is the geometry of and the movement of what is mapped.

Area Studies Tradition

The area studies tradition, like the spatial tradition, is quite strikingly represented in classical antiquity by a practitioner to whose surviving work we can point. He is Strabo, celebrated for his *Geography* which is a massive production addressed to the statesmen of Augustan Rome and intended to sum up and regularize knowledge not of the location of places and associated cartographic facts, as in the somewhat later case of Ptolemy, but of the nature of places, their character and their differentiation. Strabo exhibits interesting attributes of the area-studies tradition that can hardly be overemphasized. They are a pronounced tendency toward subscription primarily to literary standards, an almost omnivorous appetite for information and a self-conscious companionship with history.

It is an extreme good fortune to have in the ranks of modern American geography the scholar Richard Hartshorne, who has pondered the meaning of the area-studies tradition with a legal acuteness that few persons would challenge. In his *Nature of Geography*, his 1939 monograph already cited,[7] he scrutinizes exhaustively the implications of the "interesting attributes" identified in connection with Strabo, even though his concern is with quite other and much later authors, largely German. The major literary problem of unities or wholes he considers from every angle. The Gargantuan appetite for miscellaneous information he accepts and rationalizes. The companionship between area studies and history he clarifies by appraising the so-called idiographic content of both and by affirming the tie of both to what he and Sauer have called "naively given reality."

The area-studies tradition (otherwise known as the chorographic tradition) tended to be excluded from early American professional geography. Today it is beset by certain champions of the spatial tradition who would have one believe that somehow the area-studies way of organizing knowledge is only a subdepartment of spatialism. Still, area-studies as a method of presentation lives and prospers in its own right. One can turn today for reassurance on this score to practically any issue of the *Geographical Review*, just as earlier readers could turn at the opening of the century to that magazine's forerunner.

What is gained by singling out this tradition? It helps toward restoring the faith of many teachers who, being accustomed to administering learning in the area-studies style, have begun to wonder if by doing so they really were keeping in touch with professional geography. (Their doubts are owed all too much to the obscuring effect of technical words attributable to the very professionals who have been intent, ironically, upon protecting that tradition.) Among persons outside the classroom the geographer stands to gain greatly in intelligibility. The title "area-studies" itself carries an understood message in the United States today wherever there is contact with the usages of the academic community. The purpose of characterizing a place, be it neighborhood or nation-state, is readily grasped. Furthermore, recognition of the right of a geographer to be unspecialized may be expected to be forthcoming from people generally, if application for such recognition is made on the merits of this tradition, explicitly.

Man-Land Tradition

That geographers are much given to exploring man-land questions is especially evident to anyone who examines geographic output, not only in this country but also abroad. O. H. K. Spate, taking an international view, has felt justified by his observations in nominating as the most significant ancient precursor of today's geography neither Ptolemy nor Strabo nor writers typified in their outlook by the geographies of either of these two men, but rather Hippocrates, Greek physician of the 5th century B.C. who left to posterity an extended essay, *On Airs, Waters and Places*.[8] In this work made up of reflections on human health and conditions of external nature, the questions asked are such as to confine thought almost altogether to presumed influence passing from the latter to the former, questions largely about the effects of winds, drinking water and seasonal changes upon man. Understandable though this uni-directional concern may have been for Hippocrates as medical commentator, and defensible as may be the attraction that this same approach held for students of the condition of man for many, many centuries thereafter, one can only regret that this narrowed version of the man-land tradition, combining all too easily with social Darwinism of the late 19th century, practically overpowered American professional geography in the first generation of its history.[9] The premises of this version governed scores of studies by American geographers in interpreting the rise and fall of nations, the strategy of battles and the construction of public improvements. Eventually this special bias, known as environmentalism, came to be confused with the whole of the man-land tradition in the minds of many people. One can see now, looking back to the years after the ascendancy of environmentalism, that although the spatial tradition was asserting itself with varying degrees of forwardness, and that although the area-studies tradition was also making itself felt, perhaps the most interesting chapters in the story of American professional geography were being written by academicians who were reacting against environmentalism while deliberately remaining within the broad man-land tradition. The rise of culture historians during the last 30 years has meant the dropping of a curtain of culture between land and man, through which it is asserted all influence must pass. Furthermore work of both culture historians and other geographers has exhibited a reversal of the direction of the effects in Hippocrates, man appearing as an independent agent, and the land as a sufferer from action. This trend as presented in published research has reached a high point in the collection of papers titled *Man's Role in Changing the Face of the Earth*. Finally, books and articles can be called to mind that have addressed themselves to the most difficult task of all, a balanced tracing out of interaction between man and environment. Some chapters in the book mentioned above undertake just this. In fact the separateness of this approach is discerned only with difficulty in many places; however, its significance as a general research design that rises above environmentalism, while refusing to abandon the man-land tradition, cannot be mistaken.

The NCGE seems to have associated itself with the man-land tradition, from the time of founding to the present day, more than with any other tradition, although all four of the traditions are amply represented in its official magazine, *The Journal of Geography* and in the proceedings of its annual meetings. This apparent preference on the part of the NCGE members *for defining geography in terms of the man-land tradition* is strong evidence of the appeal that man-land ideas, separately stated, have for persons whose main job is teaching. It should be noted, too, that this inclination reflects a proven acceptance by the general public of learning that centers on resource use and conservation.

Earth Science Tradition

The earth science tradition, embracing study of the earth, the waters of the earth, the atmosphere surrounding the earth and the association between earth and sun, confronts one with a paradox. On the one hand one is assured by professional geographers that their participation in this tradition has declined precipitously in the course of the past few decades, while on the other one knows that college departments of geography across the nation rely substantially, for justification of their role in general education, upon curricular content springing directly from this tradition. From all the reasons that combine to account for this state of affairs, one may, by selecting only two, go far toward achieving an understanding of this tradition. First, there is the fact that American college geography, growing out of departments of geology in many crucial instances, was at one time greatly overweighted in favor of earth science, thus rendering the field unusually liable to a sense of loss as better balance came into being. (This one-time disproportion found reciprocate support for many years in the narrowed, environmentalistic interpretation of the man-land tradition.) Second, here alone in earth science does one encounter subject matter in the normal sense of the term as one reviews geographic traditions. The spatial tradition abstracts certain aspects of reality; area studies is distinguished by a point of view; the man-land tradition dwells upon relationships; but earth science is identifiable through concrete objects. Historians, sociologists and other academicians tend not only to accept but also to ask for help from this part of geography. They readily appreciate earth science as something physically associated with their subjects of study, yet generally beyond their competence to treat. From this appreciation comes strength for geography-as-earth-science in the curriculum.

Only by granting full stature to the earth science tradition can one make sense out of the oft-repeated addage, "Geography is the mother of sciences." This is the tradition that emerged in ancient Greece, most clearly in the work of Aristotle, as a wide-ranging study of natural processes in and near the surface of the earth. This is the tradition that was rejuvenated by Varenius in the 17th century as "Geographia Generalis." This is the tradition that has been subjected to subdivision as the development of science has approached the present day, yielding mineralogy, paleontology, glaciology, meterology and other specialized fields of learning.

Readers who are acquainted with American junior high schools may want to make a challenge at this point, being aware that a current revival of earth sciences is being sponsored in those schools by the field of geology. Belatedly, geography has joined in support of this revival.[10] It may be said that in this connection and in others, American professional geography may have faltered in its adherence to the earth science tradition but not given it up.

In describing geography, there would appear to be some advantages attached to isolating this final tradition. Separation improves the geographer's chances of successfully explaining to educators why geography has extreme difficulty in accommodating itself to social studies programs. Again, separate attention allows one to make understanding contact with members of the American public for whom surrounding nature is known as the geographic environment. And finally, specific reference to the geographer's earth science tradition brings into the open the basis of what is, almost without a doubt, morally the most significant concept in the entire geographic heritage, that of the earth as a unity, the single common habitat of man.

An Overview

The four traditions though distinct in logic are joined in action. One can say of geography that it pursues concurrently all four of them. Taking the traditions in varying combinations, the geographer can explain the conventional divisions of the field. Human or cultural geography turns out to consist of the first three traditions applied to human societies; physical geography, it becomes evident, is the fourth tradition prosecuted under constraints from the first and second traditions. Going further, one can uncover the meanings of "systematic geography," "regional geography," "urban geography," "industrial geography," etc.

It is to be hoped that through a widened willingness to conceive of and discuss the field in terms of these traditions, geography will be better able to secure the inner unity and outer intelligibility to which reference was made at the opening of this paper, and that thereby the effectiveness of geography's contribution to American education and to the general American welfare will be appreciably increased.

Notes

1. William Morris Davis, "An Inductive Study of the Content of Geography," *Bulletin of the American Geographical Society*, Vol. 38, No. 1 (1906), 71.
2. Richard Hartshorne, *The Nature of Geography*, Association of American Geographers (1939), and idem., *Perspective on the Nature of Geography*, Association of American Geographers (1959).
3. The essentials of several of these definitions appear in Barry N. Floyd, "Putting Geography in Its Place," *The Journal of Geography*, Vol. 62, No. 3 (March, 1963), 117–120.
4. William H. Wallace, "Freight Traffic Functions of Anglo-American Railroads," *Annals of the Association of American Geographers*, Vol. 53, No. 3 (September, 1963), 312–331.
5. Fred K. Schaefer, "Exceptionalism in Geography: A Methodological Examination," *Annals of the Association of American Geographers*, Vol. 43, No. 3 (September, 1953), 226–249.
6. James P. Latham, "Methodology for an Instrumental Geographic Analysis," *Annals of the Association of American Geographers*, Vol. 53, No. 2 (June, 1963), 194–209.
7. Hartshorne's 1959 monograph, *Perspective on the Nature of Geography*, was also cited earlier. In this later work, he responds to dissents from geographers whose preferred primary commitment lies outside the area studies tradition.
8. O. H. K. Spate, "Quantity and Quality in Geography," *Annals of the Association of American Geographers*, Vol. 50, No. 4 (December, 1960), 379.
9. Evidence of this dominance may be found in Davis's 1905 declaration: "Any statement is of geographical quality if it contains... some relation between an element of inorganic control and one of organic response" (Davis, *loc. cit.*).
10. Geography is represented on both the Steering Committee and Advisory Board of the Earth Science Curriculum Project, potentially the most influential organization acting on behalf of earth science in the schools.

The Changing Landscape of Fear

SUSAN L. CUTTER, DOUGLAS B. RICHARDSON, AND THOMAS J. WILBANKS

In the days following September 11, 2001, all geographers felt a sense of loss—people we knew perished, and along with everyone else we experienced discomfort in our own lives and a diminished level of confidence that the world will be a safe and secure place for our children and grandchildren. Many of us who are geographers felt an urge and a need to see if we could find ways to apply our knowledge and expertise to make the world more secure. A number of our colleagues assisted immediately by sharing specific geographical knowledge (such as Jack Shroder's expert knowledge on the caves in Afghanistan) or more generally by assisting rescue and relief efforts through our technical expertise in Geographic Information System (GIS) and remote sensing (such as Hunter College's Center for the Analysis and Research of Spatial Information and various geographers at federal agencies and in the private sector). Still others sought to enhance the nation's research capacity in the geographical dimensions of terrorism (the Association of American Geographers' Geographical Dimensions of Terrorism project). Many of us have given considerable thought to how our science and practice might be useful in both the short and longer terms. One result is the set of contributions to this book.

But, we fail in our social responsibility if we spend our time thinking of geography as the end. Geography is not the end; it is one of many means to the end. Our concern should be with issues and needs that transcend any one discipline. As we address issues of terrorism, utility without quality is unprofessional, but quality without utility is self-indulgent. Our challenge is to focus not on geography's general importance but on the central issues in addressing terrorism as a new reality in our lives in the United States (although, unfortunately, not a new issue in too many other parts of our world).

The September 11, 2001 events have prompted both immediate and longer-term concerns about the geographical dimensions of terrorism. Potential questions on the very nature of these types of threats, how the public perceives them, individual and societal willingness to reduce vulnerability to such threats, and ultimately our ability to manage their consequences require concerted research on the part of the geographical community, among others. Geographers are well positioned to address some of the initial questions regarding emergency management and response and some of the spatial impacts of the immediate consequences, but the research community is not sufficiently mobilized and networked internally or externally to develop a longer, sustained, and theoretically informed research agenda on the geographical dimensions of terrorism. As noted more than a decade ago, "issues of nuclear war and deterrence [and now terrorism] are inherently geographical, yet our disciplinary literature is either silent on the subject or poorly focused" (Cutter 1988: 132). Recent events provide an opportunity and a context for charting a new path to bring geographical knowledge and skills to the forefront in solving this pressing international problem.

PROMOTING LANDSCAPES OF FEAR

Terrorists (and terrorism) seek to exploit the everyday—things that people do, places that they visit, the routines of daily living, and the functioning of institutions. Terrorism is an adaptive threat which changes its target, timing, and mode of delivery as circumstances are altered. The seeming randomness of terrorist attacks (either the work of organized groups or renegade individuals) in both time and space increases public anxiety concerning terrorism. At the most fundamental level, September 11, 2001 was an attack on the two most prominent symbols of U.S. financial and military power: the World Trade Center and the Pentagon (Smith 2001, Harvey 2002). The events represented symbolic victories of chaos over order and normalcy (Alexander 2002), disruptions in and the undermining of global financial markets (Harvey 2002), a nationalization of terror (Smith 2002), and the creation of fear and uncertainty among the public, precisely the desired outcome by the perpetrators. In generating this psychological landscape of fear, people's activity patterns were and are being altered, with widespread social, political, and economic effects. The reduction in air travel by consumers in the weeks and months following September 11, 2001 was but one among many examples.

WHAT ARE THE FUNDAMENTAL ISSUES OF TERRORISM?

There are a myriad of different ways to identify and examine terrorism issues. Some of these dimensions are quite conventional, others less so. In all cases, geographical understanding provides an essential aspect of the inquiry. There are a number of dimensions of the issues that seem reasonably clear. For instance, one conventional way of looking at the topic is to distinguish four central subject-matter challenges:

1. *Reducing threats*, including a) reducing the reasons why people want to commit terrorist acts, thereby addressing root causes, and b) reducing the ability of potential terrorists to accomplish their aims, or deterrence.
2. *Detecting threats* that have not been avoided, using sensors and signature detection to spot potential actions before they happen and interrupt them.
3. *Reducing vulnerabilities to threats,* focusing on critical sectors and infrastructures, hopefully without sacrificing civil liberties and individual freedoms.
4. *Improving responses to terrorism*, emphasizing "consequence management," and also attributing causation and learning from experience (for example, forensics applied to explosive materials and anthrax strains).

A different way of viewing terrorism is according to time horizons. Immediately after September 11, 2001, governmental leaders told us that the nation was now engaged in a new "war on terrorism" that will last several years, and that our existing knowledge and technologies are needed for this war. Early estimates of the overall U.S. national effort are very large—in the range of $30 to $40 billion per year—including the formation of a new executive department, the Department of Homeland Security. Early priorities include securing national borders, supporting first responders mainly in the Federal Emergency Management Agency (FEMA) and the Department of Justice, defending against bioterrorism, and applying information technologies to improve national security.

Beyond this, we know that better knowledge and practices should be put to use in the next half decade or so, as we face a challenge that is more like a stubborn virus than a single serial killer. To address this type of need, attention often is placed on capabilities where progress can be made relatively quickly if resources are targeted carefully. Some of our CIS and GIScience tools are especially promising candidates for such enhancements, which have both positive and negative consequences (Monmonier 2002). The use of such technologies surely will help secure homelands, but at what price, the loss of personal freedoms or invasion of privacy?

There are other dimensions as well. For instance, one dimension concerns boundaries between free exchanges of information and limited ones, between classified work and unclassified work. Another differentiates between different types of threats: physical violence, chemical or biological agents, cyberterrorism, and the like. Still other themes are woven through the material that follows.

THE CHALLENGE AHEAD

The greatest challenge to geographers and our colleagues in neighboring fields of study is to stretch our minds beyond familiar research questions and specializations so as to be innovative, even ingenious, in producing new understandings that contribute to increased global security. Clearly, the most serious specific threats to security in the future will be actions that are difficult to imagine now: social concerns just beginning to bubble to the surface, technologies yet to be developed, biological agents that do not yet exist, terrorist practices that are beyond our imagination. A core challenge is to improve knowledge and institutional capacities that prepare us to deal with the unknown and the unexpected, with constant change calling for staying one step ahead instead of always being one step behind. When research requires, say, three years to produce results and another two years to communicate in print to prospective audiences, we need to be unusually prescient as we construct our research agendas related to terrorism issues, and we need to be very perceptive and skillful in convincing non-geographers that these longer-term research objectives are, in fact, truly important.

The topic of combating terrorism is not an easy one. It calls for us to stretch in directions that may be new and not altogether comfortable. It threatens to entangle us in policy agendas that many of us may consider insensitively conceived, even distasteful. It may endanger social cohesion in our own community of scholars. On the other hand, how can we turn our backs on a phenomenon that threatens political freedom, social cohesion far beyond our own cohorts, economic progress, environmental sustainability, and many other values that we hold dear, including the future security of our own children and grandchildren?

More fundamentally, geographers are not concerned only with winning the war on terrorism in the next two years or deploying new capabilities in the next five or ten. We are concerned with working toward a secure century, restoring a widespread sense of security in the global society in the longer term without undermining basic freedoms. This is the domain of the research world; assuring a stream of new knowledge, understandings, and tools for the longer term, and looking for policies and practices that—if they could be conceived and used—would make a significant difference in the quality of life.

As we prepare to create this new knowledge and understandings, what we are trying to do, in fact, is to create the new twenty-first-century utility—not a hardened infrastructure such as for power or water, but rather a geographical understanding and spatial infrastructure that helps the nation understand and respond to threats. The effort required to create this new utility to serve the nation has an historical analogy in the creation of the Tennessee Valley Authority (TVA), under Franklin Roosevelt's New Deal. The Appalachian region the southeastern United States had a long history of economic depression and was among those areas hardest hit by the Great Depression of the 1930s. The creation of the TVA, a multipurpose utility with an economic development mission, constructed dams for flood control and hydroelectric power for the region in order to: 1) bring electricity to the rural areas that did not have it; 2) stimulate new industries to promote economic development; 3) control flooding, which routinely plagued the region; and 4) develop a more sustainable and equitable future for the region's residents. This twenty-first-century utility must rely on geographical knowledge and synthesis capabilities as we begin to understand the root causes of insecurity both here and abroad, vulnerabilities and resiliencies in our daily lives and the systems that support them, and our collective role in fostering a more sustainable future, both domestically and globally.

Much of the content of this book is aimed at this longer term, and it is important for geography to join with others in the research community to assure that the long term is not neglected as research support is directed toward combating terrorism and protecting homelands in the short run. This is why the Association of American Geographers and some of its members have joined together to produce the perspectives and insights represented in this book. It is only a start, we still have a long way to go, and there are daunting intellectual and political hazards to be overcome. But if many of us will keep a part of our professional focus on this global and national issue, we have a chance to make our world better in many tangible ways.

Recreating Secure Spaces

RAY J. DEZZANI AND T. R. LAKSHMANAN

A PERSISTENT TREND IN THE MODERN ERA has been the increasing level and pace of economic, social, and cultural interactions within countries, and over time, between them. These interactions and exchanges have been made possible in recent times by the territorial state's provision (within their geography and across borders) of key public goods: law, order, "secure"interaction spaces, and physical and non-physical infrastructure networks, all of which facilitate interactions among social and economic actors (Braudel 1984, Lakshmanan 1993).

There are many recent innovations in physical infrastructure technologies (transport and communications), and in non-physical infrastructures (such as freer markets, new financial and professional management practices and international institutions) that facilitate interactions. These have helped reduce the frictions of time and distance in the control of resources across vast spaces and assisted in economic development. The level and variety of interactions make economic activities more efficient and productive and create *new* activities never before possible. This transformation appears in the form of globalization of markets and inputs, in the decentralization and networking of firms internally, and in their relationship to other firms. The resulting local networks—supplier, producer, financial, customer, technology—are leading to international networks of competition, exchange, and association, and can be viewed as the emerging dynamic "network society"(Castells 1996, 1997).

The resultant growth and development of the world economy are evident in the explosion of global trade, globally organized production systems, widespread economic growth, extensive exchange of ideas and practices, and a "borderless world". This cycle of increasing secure interaction spaces, complex and varied infrastructure networks, and the surge of technical and social innovations have indeed led to this dynamic network society with its long-term material improvements in economy and society.

Another aspect of the evolution of the dynamic global economy is that it is subject to periodic setbacks and backslidings, which occur when the quality of the key public goods noted earlier (security of interaction spaces and functionality of the transport, information, and financial networks across national and international territories) was threatened and eroded. These threats arrived in several forms: wars with territorial enemies, often deriving from security alliances and balance of power politics; and increasingly from "deterritorialized threats," ranging from transnational terrorism and proliferating weapons of mass destruction to environmental degradation and ethnic nationalism (O'Tuathail 1999).

Since contemporary transnational terrorism attempts to alter the functioning of the social fabric and bring about political change (Laqueur 1996), its actions serve to convert the basic supportive networks of our society into high risk spaces, rather than secure interactive ones. This paper examines the vulnerabilities of societal networks to terrorism and the necessary approaches to recreate secure interaction spaces.

NETWORK VULNERABILITIES FROM TERRORISM

The political events of the last decade since the dissolution of the former Soviet Union have altered the perceptions of state security to the degree that terrorism now appears to be the leading threat to national security in the early twenty-first century. The contemporary era is not unique. Terrorist threats historically have been directed at individuals, groups of individuals, or economic and military infrastructure such as the attacks upon politically elevated personages perpetrated by anarchists and budding nationalists between the late nineteenth century and the onset of the First World War. However, the nature of terrorism has changed significantly over the course of the twentieth century from anarchists and nationalists in the early part of the century; to leftist guerrillas in the middle; to right wing fundamentalist religious groups, conservative national and ethnic organizations, and anti-systematic/globalization movements in the later portion of the century (Laqueur 1996).

By definition, terrorism is purposeful only if it alters the functioning of the social fabric (Laqueur 1996). For terrorists desirous of reducing the functionality of a society, the multiplicity of physical and non-physical networks—which govern the efficient operation of national and global economies—offer crucial targets. Table 5.6.1 illustrates many of these networks and their vulnerabilities to terrorism in the form of the elevated risk to flows of goods and people, of information, and of finance.

In general, the more hierarchical the network the greater the vulnerability it sustains from terrorist threats. The hub and spoke airport system in which air traffic is heavily routed through specific hub airports, provides an example of such heightened vulnerability. In contrast, less hierarchical and more spatially distributed networks (with considerable redundancies), such as roads and the Internet experience less vulnerability (Table 5.6.1).

The continued functionality and growing productivity of these physical and non-physical networks depends upon the low and steadily declining costs of economic and social interactions within countries and among countries. Terrorist attacks, or even the threat of an attack directed against either the population or a component of infrastructure, will invoke a defensive response from the state in the form of short-term layers of infrastructure protection, which increase the costs of interaction. If territorial space and infrastructure networks are subject to penetration and destructive attack by hostile non-territorial terrorist actors, the costs of protecting the networks and providing secure spaces will have adverse affects on social interaction costs and consequently on economic and social development. Further, the increasing perceptions of personal risk impose new costs to social and economic interactions.

As the geometry of networks exhibits both links and nodes, one way of accomplishing the creation of secure spaces is by protecting nodes, which is always more economically feasible than protecting links. Depending on the network node (such as airports for passenger travel or ports for shipping), protection is an initial and necessary condition for reducing vulnerability.

RECREATING SECURE NETWORK SPACES: THE STATE'S ROLE

A primary function of the territorial state is to secure internal regions for purposes of encouraging economic and social interaction among residents. A prerequisite for this is the state's provision of law, order, and secure social interaction spaces. Such public goods provide predictability in social and economic interactions and lower private social and economic interaction costs, facilitate flows of goods and information, and enhance socioeconomic interactions. Recent efforts to deregulate and privatize economic activities from government sponsorship also have promoted the social perspective that state territories and cross-border spaces were becoming more "secure" from outside threats. The increasing security in territorial and cross-border space has meant that infrastructures—financial, logistical, and institutional networks— can develop more readily in the globally-interconnected world to promote the efficient functioning of the national and global economies.

TERRORIST THREATS TO NETWORKS

The majority of threats during the past twenty years that affected the stability of the international nation state system or the security of component networks have come from sub-state, nonterritorial entities such as ethnic or national groups seeking political autonomy or antisystematic movements seeking greater economic representation in the increasingly global economy (see chapter 3). The current threats to state security are no different. The secure spaces created by the territorial states over the past twenty years were enhanced by the collapse of the former Soviet Union and led to a decade of international economic growth and prosperity.

The majority of external threats in the past 50 years had been derived from the Cold War geopolitical order, which polarized ideologies as well as the location and direction of any threat potential. However, the risk associated with the Cold War threats were known, were "ordered"according to the existing state structure of potential conflict (nuclear or conventional), and reflected the enlightened self-interest of the state vis-à-vis their respective populations, territories, and infrastructures and reflected the goal of national survival (Brams 1975, Stein 1990, O'Tuathail 1999, Powell 1999).

Today, global dangers are viewed, in practice, as a parade of nonterritorial enemies like terrorists, rogue states, nuclear-armed agents, and the like. When nonterritorial entities are arrayed against the power of a sovereign state, a wider range of strategies might be employed directly against the population, territory, or infrastructure of the state because of the desire to equalize the great power differential that exists between the territorial state and the nonterritorial players (such as terrorists). In general, this situation creates greater uncertainty regarding the actions/strategies of nonterritorial and terrorist groups, which in turn leads to amorphous and pervasive dangers, and a greater range of risks associated with security maintenance.

The following example provides a useful case in point. A major network target of terrorism has been the transport system. The first phase of airline hijackings in the 1960s led to public disillusionment and enhanced danger associated with airline travel. However, through the institutional establishment of air marshals, terminal security checks, and other risk-mitigating strategies, air travel security was reestablished by the early 1970s. Improved security, airline deregulation, and associated technological improvements led to a dramatic expansion of air passenger traffic starting in the 1980s. Airports and transport aircraft cabins once again were perceived as secure spaces. September 11, 2001 significantly changed the secure "spaces" of air travel (airport terminals, aircraft cabins, ancillary transport system spaces), into high risk "spaces" associated with the public's perception of an enduring threat. The threat and consequently the risk, are amplified by the uncertain nature of terrorist activities directed at these likely spaces.

TABLE 5.6.1 Illustrative Network Types and Vulnerabilities

Type (stocks)	Vulnerabilities
Road	Spatially extensive, less hierarchical network with distributed links and nodes where many alternate paths are available for rerouting. Links are less vulnerable than key intersections. Risk can vary spatially with inconsistencies in the road network complexity.
Air transport	Spatially hierarchical, concentrated hub-and-spoke link and node arrangements. As a result of resource concentration spawned by the hierarchical structure limited node alternatives ensure that the nodes/airports are less secure.
Pipelines	Spatially constrained networks resulting from large capital costs produce a lack of link route alternatives. As a result, the pipelines are extremely vulnerable. However, as flows in the pipeline are dependent on the pump stations for mobility, the node structures are also vulnerable. Geographic isolation from population centers may also contribute to elevated risk.
Rail	Spatially hierarchical and constrained networks with some redundancy where link concentration and complexity is great. Links are highly vulnerable in low-density areas. Conduits from the exterior to the interior of sovereign territorial States.
Sea Lanes	Networks with spatially distributed but hierarchical nodes or, port facilities. The links are nonphysical, flexible, and determined by ships course and destination. As such, links exhibit low vulnerabilities, though ships may be vulnerable, (especially cruise ships carrying passengers). However, ships may be effectively policed and secured. Port facilities are most vulnerable. However, it is also more cost effective to secure port facilities. Effective port policing may reduce several different infrastructure network vulnerabilities.
Fiber Optic/ Telecommunications	Spatially complex distributed networks with high levels of link and node (e.g., router/switchers) redundancies. Nodes are least secure.
Cellular/ Microwave	Spatially distributed nodes only. The nodes correspond to cell towers, which are vulnerable, but if destroyed do not threaten the loss of the network. Much redundancy.
Internet	Spatially distributed links (cable, fiber optic, and nodes routers/servers). Much network link redundancy ensures that the routers/servers/nodes are the least secure structures.
Financial	Spatially hierarchical networks consisting of telecommunications links and markets. The physical market places are most vulnerable as the linkages provide routing alternatives.
Logistical	Spatially distributed organizations of multi-modal transport structures which minimize the total cost of transportation.

While airline travel is the most noticeable example, other transportation vulnerabilities include port facilities and rail infrastructure. Both of these operate under low to moderate security levels. This is especially true at the ports, where seamless intermodal freight moves (with minimal, if any inspection) to support "just-in-time" and "lean production systems." Ports also can serve as conduits for the potential movement of weapons of mass destruction.

Virtual spaces, produced by Internet linkages, are much harder to threaten (but not impossible) owing to infrastructure redundancy, in addition to high levels of security and "hardening" of certain portions of the physical network (Table 5.6.1). As such, the servers and routing apparatus are most vulnerable, and are usually associated with institutions such as government agencies, national laboratories, and colleges and universities. The Internet presents a major risk avenue into territorial states by non-territorial and terrorist actors.

Power projection by nonterritorial actors does not occur through normal diplomatic or territorial state pathways, as no traditional or established structure exists to conduct negotiations. Many territorial states, such as Israel and the United Kingdom, exercise policies that actively prohibit negotiation with nonterritorial terrorist actors. Infrastructures such as media and the Internet may become increasingly useful to both territorial states and the nonterritorial terrorist actors as a means of information signaling among the parties involved.

RECREATING SECURE NETWORKS FROM TERRORISM

The threat to states from non-state violent actors (terrorists) is not a new phenomenon and dates back to the fifteenth century with the extensive and creative use of nonstate violent actors gainfully acting in the interests of territorial states competing for prestige, territory and hegemony. From a geographical perspective, non-state actors performed tasks and exercised political options that were unavailable to the forces of traditional state actors, such as the regular armed forces. The ability to routinely violate territorial boundaries was not an option for the formal state actors. As such, non-state actors represented a minimal cost solution to territorial state governments because regular armed forces could not be employed and thus, reprisals would not be forthcoming. Similarly, non-state actors also provided a minimum risk alternative to the forces of the states because if the goal was not achieved, the state simply denied knowledge of the action. In this way, no territorial state legally violated the territory and the sovereignty of another territorial state.

The Case of Piracy

Non-state actors could be engaged for either land or naval actions. Mercenary armies roamed Europe from 1400 to 1800, and privateers scoured the seas during the same period. In many cases, when the services of these nonstate forces were no longer required, they were disbanded and scattered. However, it is not surprising that piracy greatly increased during this time period. Indeed, piracy often served the needs of weaker states against stronger states, as was the case of England in its conflict with Spain in the sixteenth and seventeenth centuries. While the names of Drake, Frobisher and Grenville were hailed with honor in England, they were decried as pirates and terrorists by the Spanish King Phillip II.

Acts of piracy involved elements of the hijacking of ships and terror enacted through the murder, rape, and torture of civilians and the poor treatment of prisoners. These activities were considered to be legal privateering raids executed by state-sanctioned, but nonstate, forces. The English privateers exploited their advantage of non-territoriality to the benefit of the English crown. Acts such as these can be described as officially sanctioned piracy (Ritchie 1986), analogous to contemporary state-sponsored terrorism. Within a short time, owing to a changing political climate between England and Spain, these same privateers had been renamed "pirates"and became an embarrassment to the English authorities who had previously exploited their services.

When it suited the interests of states to halt piracy, the patrol of the sea lanes by the navies of various cooperating states was organized. The elimination of pirates was achieved through a combination of reforms instigated by the territorial states taking action: 1) to create mechanisms that strengthened the central state; 2) against the nonterritorial means of conducting violent activities; 3); to destroy the markets where pirated goods could be traded; and 4) to improve and secure the major trade infrastructure of the day—the international sea lanes (Katele 1988, Thompson 1994).

Contemporary De-territoralized Threats

As contemporary technology—mechanical and informational— moves fast across borders accompanied by complex money flows and the availability of skilled and unskilled labor, a "borderless world"is emerging. While this borderless world is generally positive, it also holds many risks. Our vulnerability increases from terrorists and other violent actors who use our technology and informational capabilities to build black markets for weapons and launder money to support violence against our networks. Such social enemies as terrorists, nuclear outlaws, and violent fundamentalists need to be isolated, contained, and defeated.

Fundamentalism is a contemporary phenomenon, one that actively attempts to reorder society, reassert the validity of a tradition, and use traditional values in new ways in today's world with the aid of global technical and institutional means. While fundamentalism as a movement needs to be contained, it is important to recognize that it has its origins in real discontentment with the exiting world order experienced by ordinary people. Such discontentment arises from the tensions inherent in the globalization process, in the deeply disjunctive relationships among technological flows, vast money flows, and human movement across countries (Appadurai 2001), which differ significantly in levels of their physical, human, and institutional capital. While the tensions inherent in globalization process cannot be resolved completely, efforts to ameliorate them will likely address some root causes of the fundamentalist movement.

CONCLUDING COMMENTS

Our recent history is characterized by large-scale socio-economic interactions of ever increasing variety and intensity. The explosion of these interactions within and among countries is possible by the progressive increase of secure interaction spaces and the infrastructure networks that facilitate them. The outcome is the emerging "network society," a surge in technical and social innovations, and a cycle of long-term material economic growth and social development.

Periodic threats to the security of the socioeconomic networks in a society arrive in many forms, most recently from terrorism. The resulting vulnerabilities vary among network types—the more hierarchically-organized (such as a hub airport) are more vulnerable than a distributed system with redundancies (such as the internet or a road system). These vulnerabilities reduce the security of social and economic interaction spaces, raise interaction costs, and brake the economic growth and social development of the affected nation. Given its mission of providing secure interaction spaces and lowering threats to its functional network systems, the state attempts to isolate, contain, and defeat terrorist threats—sometimes using a supplementary strategy of addressing some root causes of terrorism which originated in the tensions inherent in the globalization process.

Perilous Gardens, Persistent Dreams

Healing the wounds of war and nature in Afghanistan.

BY ROB SCHULTHEIS

"To every man, his homeland is as beautiful as Kashmir."

—AFGHAN SAYING

It's barely 9 A.M. on this day in early September 2002, and already more than 300 families have arrived at the dusty row of giant tents that houses the United Nations High Commissioner for Refugees repatriation center on the outskirts of Kabul. They have come in rented buses and trucks, vehicles jammed to bursting with everything they can possibly carry. One rickety bus, adorned with dangling chains, Christmas lights, sculptures of jet fighters, and gaudy murals of lions, snowy peaks, mosques, scimitars, and peacocks, has three or four goats tethered to the roof and several hundred pounds of firewood lashed to the rear. A minibus is nearly hidden beneath a load of household furniture, carpets, bicycles, and a homemade satellite dish hammered together out of scrap metal.

Before the wars and drought, Afghanistan was a triumph of human ingenuity in one of the harshest environments on Earth.

Some of these families have been living in exile since 1979, when the Soviets invaded Afghanistan and began bombarding the country's rebellious villages into rubble. Others left during the factional fighting between rival guerrilla groups that followed the collapse of the Rabani government at the end of the '90s, and still others fled the oppression of the Taliban, which was overthrown only last year, and the ruinous drought, which has parched the country for more than four years.

At the peak of the Soviet-Afghan war, nearly half of the country's 17 million people were refugees. Close to 3 million were living in Pakistan, another million and a half in Iran, and at least 4 million were internal refugees, fleeing fighting, ethnic persecution, drought, and famine.

Now the Afghan masses are returning home. There are a half-dozen refugee resettlement centers like this one scattered around Afghanistan's frontiers, and at all of them the flow of humanity this year has been overwhelming. A month or two earlier the numbers passing through here were ten times what they are today; with winter not far off, many of the remaining refugees are postponing their repatriation until spring. But even with these reduced numbers, the system is strained to the breaking point. The wheat ration to homeward-bound families has been halved, and the amounts of cash, blankets, and rudimentary shelter-building materials per person have also shrunk.

One peasant headed for Wardak province, just west of Kabul, is loudly complaining to the UN staffers about how long it's taking to process his family's papers. (Afghans never seem to lose their capacity for outrage, impatience, and entitlement, even when their pockets and bellies are empty. *Nang*, pride: It's one of their great natural resources.) When I ask him in Dari how he is doing, he immediately stops yelling and turns to me, smiling. He visibly swells with joy as he answers: "I am going back to my village. It is the best place in the world!"

His is likely a minority view. The Afghanistan he and his fellow refugees are returning to is still a disaster zone, damaged and all but destroyed by a quarter century of savage war and the worst drought in its history. It will be a bitter homecoming for most of these happy, hopeful people: The home they dream of and long for is gone. The new Afghan government and international aid groups are just beginning to address the mine-strewn fields, ruined infrastructure, and widespread lawlessness that frustrates reconstruction efforts.

Many of the challenges are environmental: the withered crops, dwindling forests, and shrinking marshes. In Afghanistan, postwar reconstruction and environmental restoration are often the same thing.

BEFORE THE WARS AND DROUGHT, Afghanistan was a triumph of human ingenuity in one of the harshest environments on Earth. As much as 80 percent of the country's mostly rural population owned the land they lived and worked on, which inspired them to be careful stewards as well as constant innovators in folk technology and science. Old Afghanistan was a garden, ever perilous, yet rich enough to support great imperial cities like Kabul, Ghazni, Herat, and Kandahar, centers of art, learning, and culture. It was as if humans moved into Death Valley thousands of years ago and proceeded to build a series of great, prosperous, and sophisticated empires there.

Afghans adapted to life in the high deserts and dry mountains with incredible creativity. Afghanistan, for instance, is plagued by desiccating sand- and dust-bearing gales. From the seventh century on, peasants designed and built adobe horizontal-vaned windmills to harness the wind power and grind wheat into flour. These were humanity's first windmills, predating those in northern central Asia and China. Historians tell of actual wind farms consisting of 75 or more mills in a row. Windmills are still in use in many areas along the Iran-Afghanistan border.

Most amazing are the irrigation systems, called *karezes* or *qanats*. Centuries ago, Afghan villagers solved the problem of evaporation from open-air desert canals by building a system deep underground. Miracles of folk engineering and hard labor, *karezes* collect runoff, snowmelt, and groundwater high on the mountainsides and carry it down to the fertile plains, allowing agriculture to prosper in areas where it would otherwise be impossible. They can extend 20 miles or more, as deep as 100 feet beneath the surface; vertical shafts every hundred yards or so enable maintenance teams to descend each spring to repair cave-ins and clear out alluvial debris. *Karezes* are still vital to farming in vast areas of southern and eastern Afghanistan.

Other peculiarly Afghan adjustments to desert farming include fertilizer factories—giant adobe pigeon roosts that attract flocks of wild birds and collect their precious guano; solar-powered drying houses, designed to concentrate the heat of the sun and turn grapes into raisins; and pipes that channel heat from kitchen ovens beneath the floors to heat the living areas of farmhouses in winter.

Afghans' greatest cultural adaptation to a countryside too forbidding to support full-time occupation was nomadism. Contrary to their romantic image, nomads are not simply footloose people addicted to wanderlust. Through the ages, Afghan nomadic and semi-nomadic groups have put together intricate skeins of migration to keep themselves alive, and even thriving. Historically, Afghanistan's nomad tribes, now numbering more than a million people and lumped under the name "Kochis," were among the wealthiest in the country. Through their skill, the country's marginal and submarginal pasturage has supported their huge herds of transient camels, sheep, goats, cattle, and horses for thousands of years. In fact, studies by early Russian ethnographers show that the manure dropped by central Asian nomadic herds was a key factor in fertilizing the stony desert soils along migration routes. When Stalin forced the nomad tribes in Soviet Central Asia onto collective farms in the 1930s, soil quality across the area actually deteriorated.

Afghanistan is an utter disaster, and with a little luck, Afghanistan has a bright future.

I was fortunate to visit both villages and nomad camps in Afghanistan back in the 1970s and early '80s, before the war destroyed so many of them. Life there was surprisingly comfortable. In eastern Afghanistan, the typical peasant dwelling was a spacious high-walled compound with a massive multiroom adobe house inside. The rooms were decorated with murals of birds, bouquets, trees, and animals. Almost every house had a flower garden. Nomads on the move lived in tents, but most Kochi families also owned permanent dwellings. The traditional diet for "middle class" peasants was rich in butter, milk, yogurt, meat and fowl, dried fruit, and nuts. In some areas, like the inhospitable central highlands, the lifestyle was less plush, but for many rural Afghans the old way of life really worked, generation after generation: what we would call today a sustainable economic system.

> *"Be glad of a few days of peace in this garden where the nightingales sing. When the singing has been silenced the beauty of my garden will be gone, too."*
>
> —KHUSHAL KHAN KHATTAK, 17TH-CENTURY AFGHAN POET

DURING THE DECADE-LONG SOVIET OCCUPATION, VIRTUALLY the entire Afghan countryside was a free-fire zone. *Washington Post* reporter Jim Rupert surveyed war damage in eastern Afghanistan in the winter of 1985–86: Every one of the 32 villages he visited had been bombed and was in ruins. In 1986, Barnett Rubin of Helsinki Watch described "the Soviet policy of destroying all the parts of the delicate agricultural and pastoral system that undergirds food production in this semi-arid country in which agriculture is heavily dependent on irrigation. Bombing has destroyed carefully terraced hillsides... grenades and bombs have destroyed the intricate underground irrigation channels. Soviet aircraft and ground troops systematically slaughter the livestock... fragmentation bombs, artillery, and napalm have devastated carefully tended orchards and vineyards." The Soviets, mujahedin guerrilla fighters, and various warring factions also left an estimated 800,000 mines littered across the landscape, making many areas uninhabitable even after the fighting ended.

Many of Afghanistan's cities and major towns, where the Soviet occupation forces were based, escaped serious damage during the war, but in the case of the capital, Kabul, the reprieve was only temporary. In 1993–94, after the last Communist puppet regime fell, rival guerrilla groups battling for power vir-

MAP: MAX SEABOUGH SOURCE: UNITED NATIONS ENVIRONMENT PROGRAMME, HTTP://POSTCONFLICT.UNEP.CH

tually destroyed the city. Whole neighborhoods and districts were leveled in ferocious street fighting with tanks, rockets, and artillery. Most Kabulis fled. When I visited Kabul in the winter of 1996–97, the place was a city of the dead; you could stand for an hour on a major street and see only a couple of old men on creaky bicycles, a man dragging a half-ton of junk in a cart. Even today the southern and western quarters of the city resemble Hiroshima or Berlin in 1945: mile after mile of absolute annihilation in which every single building has been reduced to a gutted shell or a mound of rubble.

"May Kabul be without gold rather than without snow."
—AFGHAN SAYING

AND THEN CAME THE DROUGHT.

Afghanistan has always been a desert country. In the mid-'60s, annual precipitation in Kandahar totaled less than 8 inches; in Helmand province, in the country's southwestern corner, less than 3. (In the United States, desert is considered to begin where annual precipitation drops below 20 inches.) Yet this was enough to keep the traditional rural economy alive.

Even before the drought began, these conditions were worsening, apparently because of global warming. In 2001, the United Nation's Intergovernmental Panel on Climate Change reported a more than 4-degree temperature increase for central Asia as a whole over the last century. According to the IPCC, the 82 percent of the region marginally suited for agriculture will eventually lose from 40 to 90 percent of its productivity as a result. The panel went on: "As mountain glaciers continue to disappear [because of warming], the volume of summer runoff eventually will be reduced.... Consequences for downstream agriculture, which relies on this water for irrigation, will be unfavorable."

In 1998–99, the rains and snows in Afghanistan failed almost completely. This super-drought encompassed areas of India, Pakistan, and what was once Soviet Central Asia, but was at its most severe in Afghanistan. An estimated 5 million wheat farmers and 80,000 nomads were affected. Kochi herders brought their flocks to the city of Ghazni and began selling them at giveaway prices. The animals were going to die anyway, of hunger and thirst. When nomads sell off their breeding stock, that is the end of the road: A way of life thousands of years old was facing extinction.

Droughts are nothing new, of course, but the social mechanisms for coping with them are disappearing. When I was in the western Afghan city of Herat in 1972, a year of drought followed by heavy spring snows and floods had decimated the Kochi herds in the area. Back then, stretching the Islamic custom of *zakât*, or almsgiving, local merchants collected enough money not only to feed and house the victims but also to replenish their herds. But now the traditional safety net is unraveling, a victim of the poverty and cynicism that inevitably follow decades of violence.

The next two years brought no relief: Crop yields across Afghanistan decreased by half from 2000 to 2001, and the UN upgraded the drought to "the worst in a hundred years." In 1999–2000, 59 percent of the Afghan population had "diet surety" (they knew where their next meal was coming from) and 43 percent had access to sufficient water. But by May 2002, only 9 percent of Afghans had diet surety, and 15 percent secure water.

Not only fields are returning to desert. In 1979, 2.8 percent of Afghanistan was covered by forest, just enough to sustain small-scale logging for building material while keeping the woodlands alive. By 1996, the figure was less than 1 percent. Over that time, forests in eastern Afghanistan shrank by 90 percent. More recent reliable figures are unavailable because of war and disorder, but if anything the situation is worse. Most of the loss is from smugglers cutting the wood and shipping it into Pakistan, but the rebuilding of Afghanistan is also a factor. In Kabul today there are vast lumberyards where timber—from slender saplings to enormous forest giants—is off-loaded from trucks from the provinces. The trucks never stop coming, feeding the reconstruction of the capital.

The Afghan people know how important their forests are. In the past, villages maintained and carefully husbanded traditional forest plots nearby; now village women spend hours each day walking to the closest forests and hauling back firewood and fodder. You don't survive in the desert for thousands of years by being shortsighted or insensitive to the natural world; on the other hand, if you and your family are hungry and someone offers you cash for that precious grove of trees that serves as windbreak and fuel store for your farm, what do you do? For many, the only answer is to gamble away tomorrow to live today.

And then start planning for tomorrow again. In village after village, locals ask for roads and schools first, then agricultural assistance, and reforestation next. Several small organizations like the French MADERA and the Afghan-founded SAVE have inaugurated tree-planting programs. But the worst deforestation is taking place in areas that are also the most lawless and dangerous. In September 2002, a group of United Nations Environment Programme scientists arrived in the country to survey forest conditions. I arranged to accompany the team to Kunar province, north of the Khyber Pass in far eastern Afghanistan. Two days before we were due to leave, the trip had to be called off because of attacks on Westerners in the area by renegade warlord Gulbuddin Hekmatyar.

Forests are more than trees, of course, and along with its timber Afghanistan is at risk of losing its wolves, snow leopards, bears, and foxes, as well as alpine ruminants like Marco Polo sheep and ibex. Along Chicken Street, downtown Kabul's traditional tourist shopping bazaar, three or four shops openly sell endangered-species pelts. One place baldly calls itself the Snow Leopard Shop. Its proprietor is unapologetic: "The people in Badakhshan are happy when hunters kill these animals. The leopards and wolves kill their sheep and goats." But isn't there a danger the wild animals will be wiped out? The merchant hoots with derision. "They are everywhere in the mountains, like pests! There will always be leopards and wolves in Afghanistan!"

People used to think the same of flamingos and storks, but they are disappearing as the drought diminishes Afghanistan's rare wetlands. The lakes and swamps near Ghazni, home to Afghanistan's flamingo population, have all but vanished, as have the great Band-e-Amir lakes west of Bamiyan. The endangered Siberian cranes that visit Band-e-Amir on their annual migrations are virtually extinct: Where there were once flocks of the majestic birds, observers in 2002 found just one lone chick, wandering forlornly at the edge of the dead inland sea.

"The world lives in hope."

—AFGHAN SAYING

THE ATTITUDE OF THE SNOW LEOPARD SHOP proprietor notwithstanding, many Afghans I talked with care deeply about their vanishing wildlife, which they consider part of their national patrimony. There is an undeniable environmentalist strain in Islam (the Sermons of Hazrat Ali contain a lovely, lyrical paean to the beauties of the peacock), and Afghan poetry is rife with praise of the glories of the wild world. "In the old days, men earned fame and honor by hunting," a farmer in the Panjshir valley told me. "A hunter who killed a hundred ibex was a legend long after he died. But have you ever eaten ibex?" He made a disgusted face. "No one likes it! Tourists would like to come here and see ibex and the other wild animals, if we don't kill them all." He was one of several Afghans who asked me about the feasibility of opening up ecotourist trekking routes and guesthouses in the mountains, to bring in money. The powerful Afghan vice president, Karim Khalili, head of the Hazara tribe, was already talking about national parks and tourism when I interviewed him back in 1996, when he and his people were fighting for their lives against the Taliban and Al Qaeda.

You don't survive in the desert for thousands of years by being shortsighted or insensitive to the natural world.

Afghanistan is unlikely to host a Club Med anytime soon, but given a modicum of support, the Afghans could turn their wilderness and wildlife into economic assets. After all, tourism was the country's biggest source of hard currency before the wars began in 1979. On my 2002 flight from the United Arab Emirates into Kabul, I encountered a group of seven elderly but intrepid European women, off on a two-week tour of Kabul and Mazar-e-Sharif. In Afghanistan, improbability is the order of the day.

IN THE COURSE OF MY TRAVELS AROUND THE COUNTRY, I came to two paradoxical conclusions: Afghanistan is an utter disaster, and with a little luck, Afghanistan has a bright future. The two perceptions kept shuffling back and forth, sometimes occurring simultaneously.

In Kabul, for instance, the drought is so severe that the Kabul River, normally a mini-Danube rushing through the heart of the city, has dried into a stagnant marshy ditch. In one area near the Old City, enterprising merchants looking for rent-free real estate have built an entire bazaar in what was once the river bottom. "In my neighborhood, the water table has fallen six and a half meters in the last year," a young Kabuli doctor told me. Neighbors have to pool their resources to drive their communal wells deeper into the aquifers beneath the city, or walk to the nearest functioning well and carry water home all day. Everywhere I saw men, women, and children hauling jerry cans and pails of water block after block in the dust and the heat.

And yet, the city is rapidly rebuilding. In the few months after the Taliban's overthrow, Kabul's population soared to more than 3 million, with returnees camping out in ruins and abandoned buildings while they put up new homes. Along the main north-south thoroughfare in eastern Kabul there is a neighborhood a mile long devoted entirely to building construction. Workshops clatter and bang away all night, cutting rebar. New businesses are starting up everywhere: In one bazaar, I even found a row of three stalls selling flowering plants, shrubs, and saplings for home gardeners. Though the country is almost devoid of telephone service, and electricity itself is spotty, everyone was talking about the new global Internet economy, and computer schools were springing up to serve it. I interviewed scores of Afghans from all of the nation's varied ethnic groups, Pashtun, Tajik, Hazara, Aimak, and Uzbek, and every single one of them was focused on the future, not the past. Ex-guerrilla fighters talked of reviving the family farm; cab drivers' sons planned to start trucking companies, or become doctors; schoolgirls dreamed of going to university.

"Even the highest mountain has a trail to the top."

—AFGHAN SAYING

HERE'S WHAT GIVES ME SUCH HOPE FOR AFGHANISTAN: the Saga of the Helicopter Gunship. In the early 1980s, the mujahedin northeast of Kabul managed to shoot down a Soviet Mi-24 Hind gunship, the most fearsome weapon in the Red Army's arsenal. (A single Mi-24 packed 168 small air-to-ground rockets, a rotary cannon that fired ten rounds per second, and two thousand-pound bombs.) The helicopter crashed close to a major foot trail, and it instantly became so famous a landmark that an enterprising local opened a teahouse in its fuselage. The jury-rigged *chaikhanna* soon became a popular stopover for travelers: Guerrillas, traders, nomads, and refugees relished resting in the shelter of what was once an object of dread and terror, sipping tea, dining, and sleeping overnight before hitting the trail again.

The farmer drops a hefty cobblestone into the darkness. A second or two later we hear a deep splash. The old man smiles.

And then one day it was gone. A British journalist returned after a year's absence to find the spot empty, the massive metal carapace vanished without a trace. When he asked his Afghan hosts what had happened to it, they led him down to the nearest dirt road. There was the gunship, now outfitted with axles, tires, and the motor and transmission from a junked Soviet jeep, rolling along with a full load of passengers and a mountain of luggage lashed on top. Adapting to changing markets, the teahouse owner had turned the helicopter into a minibus and was hauling people and freight from village to village around the foothills.

When I visited the Panjshir valley northeast of Kabul, I found the villages were all lighted at night. A UN or Afghan government electrification program? No. It turned out that years ago a local villager had read a book on generators, and designed and tinkered together his own pocket hydro plant. Other villagers copied it, and now the whole valley is powered by a series of small, homegrown hydro projects.

North of Kabul, on the fertile Shomali Plains, mine-clearing is taking place at breakneck speed. Everywhere you look, Halo Trust de-miners in heavy flak vests and plastic visors are crouched over antipersonnel and anti-tank mines, booby-trapped shells and aircraft bombs, cutting wires and defusing. The moment a farmhouse and its fields are cleared, the family moves back in and starts repairing and planting. Hundreds of families are camped out at the edge of the mined danger zone, waiting for the chance to go home. "My family has lived here for hundreds of years," one ruddy-faced farmer tells me as his children swarm laughing around him. "I had to leave when the Taliban came. We lived in Pakistan for six years. It was like hell. When we get home, we will never leave again." He and his fellow villagers are already talking about building a new school, to ensure the community's future.

Afghanistan's greatest natural resource isn't gas or oil, or copper, or rubies and emeralds: It is the Afghans themselves. If this desperate, damaged country eventually recovers, it will be because of the people, their intimate knowledge of their uncompromising homeland, and their passionate love for it. "Give Afghanistan two or three years of peace, and a year or two of normal rainfall, and it will be back on track," an American diplomat told me. "Bet on it." In the winter of 2003, heavy snows finally returned to the mountains, giving hope of a break in the long drought.

At the edge of Shomali, at a place called Deh Sabz, another elderly refugee farmer visits what was once his home. The land has dried out: Unlike the central Shomali, farming here was dependent on irrigation, and the Taliban dynamited the *karezes* that brought water from the mountains. Together, we peer down the maintenance shafts that lead to the canal. We can't see the bottom. The farmer picks up a hefty cobblestone and drops it into the darkness. A second or two later we hear a deep splash. The old man smiles. He will return with family and friends to clear the system downstream and let the water run onto his fields again.

He walks down to where a trickle of water still emerges from the tunnel mouth, and carefully, gently, he scours the built-up sand and mud away from the opening. The rivulet becomes a streamlet, and the desert silence is suddenly broken by the silver music of water.

ON THE WEB *For Beth Wald's remarkable images of Afghanistan, see the photo gallery at* www.sierraclub.org/sierra/afghanistan.

ROB SCHULTHEIS *has visited Afghanistan over 30 times since 1972. He has reported on events there for* Time, CBS, *NPR, the* New York Times Magazine *and* Smithsonian.

AFTER APARTHEID ...

Change is everywhere in South Africa

Judith Fein

"He's our Moses," said a middle-aged black South African school teacher when my husband and I asked him how he felt about Nelson Mandela. "He led us out of the slavery of apartheid, through the Reed Sea (once called the Red Sea) of uncertainty and into the promised land of a democratic South Africa. It's only 10 years since apartheid officially ended. And you cannot imagine the change."

Change. It's exciting. It's vibrant. It's everywhere in South Africa, as people try to adapt and redefine themselves in relation to the new reality.

The black majority, despite huge problems like unemployment and AIDS, is walking easily through doors that were shut to them for decades. The white community—which had both supported and condemned apartheid after it was instituted in 1948—has been forced to do some soul searching and to accept the fact that it is no longer in charge.

In many cases today, members of the white community are last in line for plum jobs and government favors, and they are—excuse the expression—sweating through the changes.

In Johannesburg, beautiful, brainy, black tour operator Thuli Khumalo is multilingual (English, her native Zulu, German, Xhosa, Sotho and Tswana) and she whisks visitors through areas that either didn't exist or would have been off-limits a decade ago.

"Come, come, sisi (sister)," she said with her characteristic friendliness. "I am going to take you to our Apartheid Museum. It will teach you about what we went through."

Outside the entry, I sat for a moment on a bench and asked Thuli to join me.

"I can't, sisi," she said. "This bench is not for me."

I twisted around and saw a sign on the bench: "Europeans Only." Repulsed, I jumped up.

The entry to the museum was equally painful. Each of us, regardless of our color, was given a ticket stamped "white" or "non-white," and this determined which door we could use. My husband was white. I was nonwhite, separated from the whites. We were thrown into a simulation of the apartheid experience.

Inside the museum, there was little noise or chit-chat. Visitors walked down a hallway lined with the humiliating and hated passbooks that blacks had to carry for identity checks. Without them, they could be summarily arrested and jailed. Walls were lined with photos and documents about racial classification and bizarre laws governing who was deemed black or colored.

(A total of 2,823 people did the "chameleon dance" and had their color changed by the stroke of a government pen after they appealed their classification. In the year 1986 alone, 702 coloreds became white, 19 whites became colored, 249 blacks became colored and one Indian became white.)

In succeeding rooms, video screens played footage of Afrikaans nationalist propaganda, police beatings, black resistance and interviews with Nelson Mandela, Winnie Mandela and murdered leader Steve Biko. Rooms were filled with jail cells, towering armored tanks and nooses used for lynchings. The corridors grew narrower, darker, the noise of screeching sirens increased, the videos multiplied, and the volume was cranked up until it became almost unbearable.

The small group of German tourists I was with exited the museum, saying they were exhausted. One man insisted he was overwhelmed. "That's what it was like for us, *boet* (brother)," said Thuli.

The next stop was at Gold Reef, one of the oddest pairings of tourism and tragedy I have experienced anywhere. Old gold mines provided wealth in Johannesburg but were the source of misery to the blacks, who were separated from their families and had to work there for a pittance.

Today, they have been turned into a Disneyland-like theme park. The old mine cars and equipment and the mine manager's house and administrative buildings have been transformed into rides and shops. Only when we were given hard hats and led into the bowels of the earth did we have a sense of the claustrophobic and brutal reality of the 28,000 mine workers.

The guide, a Chinese man deemed colored during apartheid, was upbeat and funny, and he made sure everyone had a grand time. He also confessed that guides were not allowed to give statistics about how many people died in the mines or other grisly details.

Anyone who remembers apartheid recalls Soweto (it means South West township) and the violence that erupted there as blacks rose to resist their oppressors.

Today, visitors are welcomed to Soweto, and Thuli drove us past the house where her grandparents lived, took us to the central market (which has everything from sandals made from tires to bowls crafted from telephone wires to beauty potions) and led us into Wandie's restaurant to sample delicious South African favorites like *pap* (corn porridge), *ting* (sorghum), *idombolo* (dumplings), *boerwoers* (beef sausages) and *gwinya* (fat cakes, an Afrikaans treat that has been fried in fat and has a sweet or savory filling).

After lunch, we rode by Mandela's house, his ex-wife Winnie's palatial residence and the square where 16-year-old Hector Peterson was shot by police in 1976 during a peaceful student demonstration. Then we visited a few of the houses and saw the shameful living conditions of some, the upgraded residences of others and the middle-class comfort of those who were upwardly mobile.

"Want to visit a *shebeen*?" a friendly Soweto boy asked us. "It's like a speakeasy." We followed him into his neighbor's small living room, which had been turned into a de facto bar. Music blared and unemployed locals hung out and drank beer, served ice cold by the woman of the house.

Back in the days of apartheid, *shebeens* served the same purpose—drinking, music and dancing—but were also meeting places to foment revolution. The police often raided the *shebeens* in a futile attempt to staunch those efforts.

One creative solution to unemployment is "unofficial" selling—along the principal streets and at busy intersections in Joburg. Artists and craftsmen ply their wares—and tourists can buy sculptures, jewelry, decorative items and paintings directly from the artists. Bargaining is de rigueur, and it is not unusual for vendors to settle for 30 percent, 40 percent or even 50 percent of their original asking price.

On weekends, the official Bruma flea market offers bargains on everything from clothes to art and souvenirs. The quality is uneven, but the prices are generally lower than the Rosebank Rooftop market, which is shoppers' heaven on Sundays. The spectacular handicrafts come from all over Africa.

For more adventurous art hounds, a three-hour trip from Joburg to the Ndebele tribe is well worth the visit. The women of the tribe paint their houses, inside and out, according to their inspirations. The brightly colored, geometrically patterned abodes are world famous.

Queen among the painters is Esther Mahlangu, who teaches the secrets of Ndebele art to children under a tree. Esther's art—painting and beadwork—is gobbled up by collectors, but if you are lucky, she will have pieces on hand to sell—and Thuli can act as translator because Esther's native Zulu is close to the Ndebele language.

If you can't buy from Esther, other Ndebele women sell beadwork and paintings in front of their houses at reasonable prices. The end of apartheid has opened up tribal villages to tourism for a much-needed infusion of capital.

In Cape Town, our guide, Tariek, drove past St. George's Cathedral, where Anglican Bishop Desmond Tutu had his congregation. During apartheid, guides said, the police hounded Tutu, arrested him and placed him under curfew, and his family was subjected to "humiliation." When apartheid was over, Tutu became the first black archbishop, and today he is revered as a hero.

Then Tariek took us to the area where he used to live and, growing teary-eyed, recounted how his home was razed and his family displaced in a forced removal—a fate 60,000 other black people in his district had to endure.

He was proud that he and the other coloreds took part in marches as part of their organized defiance. When I asked Tariek if he was angry, he said he didn't like to look backward. He was glad he could work—being a tour guide would have been impossible for a colored man during apartheid—and although the pace of economic growth wasn't always fast enough, it was, at last, happening.

"Black kids of today are not interested in apartheid. It's like our parents talking to us about World War II. ... They never knew what it was like before. All our kids today live in a bubble, with no sense of the past."

JOHNNY CLEGG
rocker known as the "white Zulu" who was active in the anti-aparthied movement

---

The Waterfront area of Cape Town offers 380 shops, 72 restaurants and eateries, seven hotels and the opportunity to board a boat for the seven-mile pilgrimage to Robben Island, where Nelson Mandela was held prisoner for 18 years. Buses take visitors to the bleakest part of the island—torrid in summer and freezing in winter—where the prisoners worked the limestone quarries with pick axes until the dust and the glare of the sun on the stone damaged their eyes. Then the tourists were deposited at the entry to the harsh prison, where their guide turned out to be a former political prisoner.

Our guide, Patrick Matanjana, spoke with acceptance and without bitterness about his incarceration, but admitted he took the job because he needed money, not because he wanted to come back to the locus of his suffering. He walked us through the washrooms, the yard and the small, sparse cell where Mandela was incarcerated. It felt like horrible but hallowed ground. It was here the politics of the new South Africa were forged by the modest, unassuming icon of the revolution.

Many people I met in Cape Town—blacks and whites alike—told me their memories of the day Nelson Mandela and other prisoners were released from Robben Island. The nation held its breath, because after a total of 27 years of incarceration, Mandela walked out a free man.

Many whites had fled the country (several black people I met referred to this as the "chicken run"), fearing a bloodbath. But in one of the miracles of the last century, the man whose spirit they tried to break came out of confinement and preached not revenge and violence but love and brotherhood. He became—and still is—a model for all humanity.

In Cape Town, a cab driver took us to Langa, another township. Unlike Soweto, the Langa we saw had no comfy homes with modern conveniences. It was subsistence living, with few amenities and crowded conditions. But Langa is using a lot of homegrown talent and ingenuity to compensate for economic hardship. It has a craft and gift shop, and locals run walking tours of the township.

"We had visitors coming through in cars, staring at us," said Shelly, our young guide. "We felt like they were on safari, and we were being stalked like animals. So we decided to do the tours ourselves."

Shelly spoke about her history, her family and what it was like growing up there. "The hardest thing about apartheid for me, personally, was that I never got to know my dad," she said. "You see, he was white, and so he wasn't allowed here. For this stupid reason, I could never see him, and he was a stranger to me. This has influenced all my life."

At the intimate and elegant Cape Grace Hotel, in Cape Town, black culture is now incorporated into the activities, the menu and even the spa treatments. The signature African Cape massage is based on the circular dance movements of the Khoi-San tribe. The massage medium includes shea butter, among other things.

In the Durban area, the end of apartheid meant native tribes could create employment by providing services to tourists. At Phe Zulu, visitors can experience Zulu traditions in a re-created village. I have heard that other villages are more touristy and found this one to be intimate, friendly, informative and rather delightful.

Young men in warrior garb perform vigorous dances and go through courting rituals with traditionally dressed young women.

A guide named Washington leads visitors into a grass hut and tells them how to make Zulu beer (from maize meal, sorghum malt and maize malt), what the colors in a beaded Zulu love letter mean (red means "my heart bleeds for you," blue means, "my love is endless like the sky" and black means "I'm as black as the rafters in the house from missing you") and about the *lobola*, or bride price, that a suitor must pay (11 cows).

At the highly recommended Phinda game reserve, democracy means blacks work alongside whites as managers, trackers, guides and security guards.

It is not only an awesome experience to find baby elephants frolicking in a lake, a pride of lions feeding on a freshly killed zebra or a leopard darting through the bush at night, but it is also soul-stirring to sit around a campfire where a chef from Zimbabwe prepares food in a pit made from termite mounds mixed with termite saliva and hear tribal stories from the heart of Africa.

Before leaving South Africa, I spent some time with Johnny Clegg, the famed rocker who is referred to as the "white Zulu."

He was active in the antiapartheid movement, put his life on the line, militated for the miners and is considered a revolutionary brother by every black South African I spoke to.

What he said about young South Africans was arresting: "Black kids of today are not interested in apartheid. It's like our parents talking to us about World War II. They are aware of brands and the global youth culture. You tell them about the progress that has been made, but they can't relate. They never knew what it was like before. All our kids today live in a bubble, with no sense of the past. They are also worried about AIDS, drugs, the new branded products they are bombarded with. Apartheid isn't on their radar."

It certainly was on the radar of all the grown-up South Africans I met, and it was on my mind every day I spent in the country.

The movement from oppression to freedom has created a society that is vigorous, exciting and full of passion, surprises and opportunity. The emotions of the people who have lived through apartheid and are willing to speak about it make a visit there unlike a visit to any other country: Go before it becomes a story in the history books.

If you go ...

For general South Africa tourism information: www.southafrica.net

- South Africa Airways, which offers nonstop flights and has many special deals: www.flysaa.com
- The Cape Grace in Cape Town, one of the world's best boutique hotels: www.capegrace.com
- Phinda, part of the Conservation Corporation Africa group, which operates 30 camps and 15 lodges throughout Africa: www.ccafrica.com
- Wandie's restaurant in Soweto: www.guides.com/detail.cfm?detailID=187649
- Apartheid Museum: www.apartheidmuseum.org
- Gold Reef: goldreefcity.co.za
- Robben Island: www.freedom.co.za
- Phe Zulu: info@tourism-kzn.or
- Thuli Khumalo, a Zulu tour operator who arranges customized visits: ATAMELA@webmail.co.za

UNIT 2

Human-Environment Relations

Unit Selections

Key Points to Consider

- What are the long-range implications of atmospheric pollution? Explain the greenhouse effect.
- How can the problem of regional transfer of pollutants be solved?
- The manufacture of goods needed by humans produces pollutants that degrade the environment. How can this dilemma be solved? Does the Indonesian logging plan have the answer?
- Where in the world are there serious problems of desertification and drought? Why are these areas increasing in size?
- How are you, as an individual, related to the land? Does urban sprawl concern you? Explain.
- Are the oceans in danger of being over-fished and polluted? How can the oceans be protected?

Links: www.dushkin.com/online/
These sites are annotated in the World Wide Web pages.

Alliance for Global Sustainability (AGS)
http://www.global-sustainability.org

Human Geography
http://www.geog.le.ac.uk/cti/hum.html

The North-South Institute
http://www.nsi-ins.ca/ensi/index.html

United Nations Environment Programme (UNEP)
http://www.unep.ch

US Global Change Research Program
http://www.usgcrp.gov

World Health Organization
http://www.who.int

The home of humankind is Earth's surface and the thin layer of atmosphere enveloping it. Here the human populace has struggled over time to change the physical setting and to create the telltale signs of occupation. Humankind has greatly modified Earth's surface to suit its purposes. At the same time, we have been greatly influenced by the very environment that we have worked to change.

This basic relationship of humans and land is important in geography and, in unit 1, William Pattison identified it as one of the four traditions of geography. Geographers observe, study, and analyze the ways in which human occupants of Earth have interacted with the physical environment. This unit presents a number of articles that illustrate the theme of human-environment relationships. In some cases, the association of humans and the physical world has been mutually beneficial; in others, environmental degradation has been the result.

At the present time, the potential for major modifications of Earth's surface and atmosphere is greater than at any other time in history. It is crucial that the environmental consequences of these modifications be clearly understood before such efforts are undertaken.

The first article in this unit illustrates how cities create their own weather. A plan to harvest logs in an Indonesian forest without degrading the environment is discussed next. Two short articles follow with speculations on ways to reduce carbon dioxide accumulations in the atmosphere. In the next article, states and municipalities find novel ways to save water and avoid pollution. The final article suggests that every state in the U.S. influences the condition of the oceans.

This unit provides a small sample of the many ways in which humans interact with the environment. The outcomes of these interactions may be positive or negative. They may enhance the position of humankind and protect the environment, or they may do just the opposite. We human beings are the guardians of the physical world. We have it in our power to protect, to neglect, or to destroy.

How Cities Make Their Own Weather

WHEN HOUSTON IS HIT BY A SUDDEN storm, the city may be partly to blame. Increasingly, urban centers don't merely endure bad weather; they help create it. Researchers believe the phenomenon may be more common now than ever before.

Scientists have known for 200 years that the temperature in a city can be higher than that in its environs—something they learned when an amateur weather watcher detected a 1.58° F temperature difference between London and its suburbs. Modern cities, with their cars and heat-trapping buildings, can create an even bigger temperature gap, sometimes as much as 10° F.

Islands of urban heat can do funny things with weather. Hot city air, like hot air anywhere else, rises—even more so because of the turbulence caused by tall buildings. When that air is damp enough and collides with colder layers above it, water can condense out as a sudden burst of rain, especially if there are few frontal systems to disrupt the layers, as in summer. In a spot storm above a city or just downwind to it, it's likely that nature alone isn't behind the downpour.

NASA and the University of Arkansas have been using satellite mapping and ground-based temperature readings to determine how widespread this phenomenon is. This spring researchers got a surprise when they turned their attention to Houston. Because it's near a coast and sea breezes tend to cool and disperse hot air, Houston was thought to be comparatively safe from homemade rain. Now it appears that the opposite may be true. "The sea breeze may exacerbate the rainfall," says research meteorologist Marshall Shepherd of NASA's Goddard Space Flight Center. The warm air and sea air collide, he explains, and "move straight up like the front ends of two cars that hit head on, providing a pump of moist air that helps thunderstorms develop."

Hot, waterlogged cities can be cooled off in the usual ways—by limiting auto exhaust, for example. Using light-colored roofing and paving materials in place of black, heat-absorbing tar will also help. As a bonus, the cooler roof will reduce the need for air conditioning.

—By Jeffrey Kluger

How Cities Make Their Own Weather

1 In big cities, heat-absorbing roofs, blacktop pavement and auto exhaust trap the sun's rays and warm the air

2 Late in the day, the accumulated heat starts to be released. The lighter, warmed air begins to rise

3 In cities near large bodies of water, moist air flows in toward the rising urban air

4 The moist air and warm city air collide and drive each other higher, hitting the cooler layer of air above and creating clouds and rain

5 The prevailing wind blows the clouds. Areas downwind of cities get more rain than those upwind

The Race To Save A Rainforest

Can an experiment in Indonesia prove the merits of sustainable logging to Big Timber?

THE INDONESIAN VILLAGE of Long Pai lies a grueling three-hour drive on rutted dirt roads from the provincial capital of Tanjung Redeb. It is a river town in Borneo that is little changed from a century ago when it served as the backdrop for Joseph Conrad's *Lord Jim.* Agus Heryanto, a local representative of the Nature Conservancy, a Virginia-based environmental group, has made the bone-rattling trip to convince villagers that they should embrace a plan to log the lush forests nearby. By the dim light of a single bulb in a stilt hut, Agus tells village elders about a trust fund supported by logging revenues that will pay for roads, medical clinics, and electric generators. And, promises Agus, the logging will be done in an environmentally responsible way.

The villagers listen to the pitch, but remain suspicious. Three years earlier, they drove out the company that wants to do the logging—Sumalindo Lestari Jaya—after disputes over payments and improper cutting of valuable trees. Agus says that this time the trust will protect against wrongdoing. Finally, at a second meeting four days later, the Long Pai elders agree to the plan. By yearend they hope to have a formal contract.

There seems to be a disconnect here. Indonesia is known for the rampant destruction of its rainforest. Now one of the country's most powerful logging companies allies itself with a Western environmental group to make nice with impoverished villagers? What happened in that stilt hut is an exampe of globalization at work—an intricate dance involving Western consumers, multinational companies, environmental activists, native forest dwellers, and local conglomerates. The aim is to combine the profit motive with the need to end the shocking degradation of the forest. A long shot perhaps, but some action is desperately needed, and it is starting in Borneo.

REWARD SYSTEM

THE SUMALINDO DEAL was indirectly driven by the vigorous campaigning of environmental groups. They have put pressure on Western retailers such as Home Depot (HD), Ikea, and Kinko's to stop buying wood, pulp, and paper from companies that do not follow sustainable practices. The retailers in turn are leery of infuriating shoppers appalled by the rape of the forests. "We can give logging companies an incentive to manage the forest," says Nigel Sizer, director of the Asia-Pacific Forests program at the Nature Conservancy. The effort is already showing results: Sumalindo, afraid of losing access to the all-important U.S. and European markets, is working with the Nature Conservancy to clean up the way it cuts timber.

The initiative still has a long way to go. Sumalindo is one of only a handful of enterprises taking even modest steps toward low-impact logging. Still, if the Indonesian effort is successful, it could serve as a template for saving forests in Southeast Asia, the Amazon basin, and Siberia.

BAR CODES VS. BOYCOTTS

BORNEO'S FORESTS don't have much time. Many of the island's great stands of trees have been felled, and Indonesian loggers continue to decimate an area half the size of Switzerland each year. Environmentalists warn that Borneo's wild orangutans, sun bears, and clouded leopards could be wiped out in 10 to 20 years, as well as countless other species. Plus, thousands of loggers could end up jobless in a land so scarred that it may be nearly impossible to revive it for other commercial uses.

Environmentalists hope to drive many more efforts like Sumalindo's. In a project set to begin before yearend, the Nature Conservancy has teamed up with Home Depot Inc. and a British aid agency to track lumber from the stump to the store. Logs and boards will be tagged with bar codes that allow buyers to determine whether their wood was harvested in a sustainable manner. "People were asking what we were doing not to add to the deforestation of the world," says Ron Jarvis, head of lumber merchandising at Home Depot. "We didn't have the answer."

Where Home Depot and the Nature Conservancy are offering carrots, more radical environmentalists are wielding sticks. Rainforest Action Network and Greenpeace International want consumers to stop buying Indonesian timber products. They think the scale of illegal Indonesian logging is so vast and the legal concessions so destructive that reforms like those proposed by the Nature Conservancy are doomed. Michael Brune, executive director of San Francisco-based Rainforest Action, says his group has pushed for a boycott of Indonesian timber products for two years. "This is not a time to establish small toeholds of good production."

The protests are having an effect. Home Depot has cut its purchases of lumber from Indonesia by more than three-quarters since 2000. In early 2002, the American Forest & Paper Assn. trade group said its members would stop buying illegal wood products, singling out Indonesia as a special concern. And Ikea requires that all of its tropical hardwoods be cut in areas okayed by the Forest Stewardship Council, a German group that issues certificates to companies following sustainable practices. Ikea cut off Indonesian teak purchases in October, 2001, when battles between local villagers and a government-owned supplier in Java over ownership of the trees short-circuited the certification process. "We regret that it happened," says Ikea forestry coordinator Par Stenmark. "But it shows that the Forest Stewardship Council certification system is working."

Pressure from Western shoppers is felt in the forest

A boycott is just the outcome Sumalindo wants to avoid—because the corporate damage could go far beyond Sumalindo itself. The company is owned by one of Indonesia's largest wood empires, the Hasko Group.

Consequently, Sumalindo is making an effort to change. In a timber concession 200 kilometers south of Long Pai, it has worked with German and U.S. experts to minimize damage to the forest and is aiming for certification from the Forest Stewardship Council. That's a sacrifice, given that the three-week evaluation costs some $50,000. Other costs may ultimately prove much higher: Sumalindo had to agree not to cut some 212,000 acres—nearly a third of one of its concessions.

Sumalindo is also paying for a variety of measures to lessen its environmental impact. These include a $500,000 aerial tram that lifts logs out of steep valleys rather than pulling them through the forest, which can cause erosion. Nature Conservancy experts also teach loggers how to fell a tree so that it ends up closer to a road, minimizing damage to the forest floor by reducing dragging. And they counsel against slashing through the forest's thick underbrush with a bulldozer.

SLASHERS AND POACHERS

THE DIFFERENCE between Sumalindo's methods and those of other companies—especially those that are government-owned—is stark. Just beyond the borders of Sumalindo's land in one of Borneo's most important watersheds lies a concession owned by the government logging company, PT Inhutani. Much of the acreage has already been cut. Gulleys full of eroded tropical soil abut the road. Wildfires raged across huge swaths of the land five years ago, after intensive logging. The company declined to comment.

Even worse are the poachers. These illegal loggers cut with impunity, often working in plain sight along main roads. They can take down a 100-year-old tree in under 10 minutes. On the banks of the Segah River, which runs through the area, some 300 logs waiting for shipment lacked any identifying marks—a telltale sign of illicit cutting. Most of this wood is smuggled across the border to Malaysia, where it can be imported with a small bribe, according to Indonesian officials and environmental groups. A senior Malaysian official, saying his country takes smuggling "very seriously," notes that Malaysia has banned imports of some wood from Indonesia. Much, though, still gets through, and eventually finds its way to mills in China. "Illegal logging is our responsibility," says Wahjudi Wardojo, the highest ranking civil servant at Indonesia's Forestry Ministry. "But without support from other countries, it's useless."

So illegal logging continues—and even most of the legal logging in Indonesia is not done sustainably. If contracts such as the proposed pact between Sumalindo and Long Pai become the norm, villagers will have a bottom-line incentive to see that all logs that are cut are accounted for, and the tracking system should ensure that the timber can be traced. But if other companies—and countries—don't sign on, the effort is doomed. And it remains to be seen if Western consumers will pay a premium for wood harvested sustainably. "We are struggling," says Wahjudi. "Our time is very limited." If they don't succeed, another corner of the earth's forests will be gone forever.

By Mark L. Clifford in Long Pai, Indonesia, with Hiroko Tashiro in Tokyo and Anand Natarajan in Atlanta

Texas and Water

Pay up or dry up

Another state trying not to go thirsty

Austin

OIL built Texas, but water will shape its future. With five droughts in the past four years, water is running out fast. The Rio Grande, one of the state's main sources, has failed to reach the Gulf of Mexico for the first time in about 50 years. And demand is soaring, with the number of Texans expected to double by 2050. If this goes on, El Paso will run out of water in 20 years and other cities not much later.

Can the supply be increased? Texas has been squabbling with its neighbours about the division of the Rio Grande's water. There is talk of new reservoirs, but they may not be ready before El Paso and other cities have gone dry. Since surface water is so scarce, many people want to pump more water from underground aquifers. This would help—55% of the state's population already depends on such water for drinking and agriculture—but reliance on well water, if not handled carefully, could create vast environmental problems and leave the state with no water reserves.

So Texas is again wondering how to cut demand. Nobody can stop the state's population growing. But there are surely ways to limit the amount of water each Texan uses. Americans consume more water per head than most people, yet the price (a third of a cent per gallon) is lower than in most other rich countries. Water takes up less than 0.8% of the average American household budget. In theory, a price increase could achieve two things. It could raise money to help build better pipelines. It could also tame consumption, perhaps by as much as 20% in some areas.

Meanwhile, however, T. Boone Pickens, an erstwhile corporate raider, has been working on the supply side. He has been busy buying water rights in western Texas, setting up a consortium of landowners who have agreed to sell him their water at a given price. Mr Pickens hopes to pipe 150,000-200,000 acre-feet of water from the vast Ogallala aquifer in the Texas Panhandle to the state's parched cities. (An acre-foot, enough water to cover an acre a foot deep, is 326,000 gallons, or some 1.2m litres.)

The cities would pay on a sliding scale, depending on the distance Mr Pickens has to pump the stuff. Far-away El Paso would pay $1,400 per acre-foot, Dallas around $800. Much of the projected revenue of $200m would go into building pipelines to carry the stuff. But these are still stiff prices: the state of California recently paid $260 an acre-foot in a similar deal.

Environmentalists have reason to worry, too. The Ogallala replenishes itself at a rate of less than one acre-foot a year. In these circumstances water extraction is little different from mining. Indeed, Texas law treats water much as it does oil or gold: anybody who has access to the stuff can extract as much as he wants.

In theory, such water is protected by Texas's Groundwater Conservation Districts. But the state's strong tradition of property rights limits the GCDs' ability to control men like Mr Pickens. C.E. Williams, general manager of the Panhandle's Groundwater Conservation District No. 3, which deals with Mr Pickens, thinks the current crisis justifies pumping up to 50% of the Ogallala. But Mr Pickens's methods, he argues, would empty it within 25 years.

Mr Williams is pinning his hopes on a bill that is currently being discussed in the state legislature. This would allow more GCDs to be established, and empower them to levy a fee of at least two-and-a-half cents per 1,000 gallons to study the effects of pumping and set up replenishment projects. The only trouble is, Mr Pickens may have already got himself exempted from the fee.

Environmental enemy No. 1

Cleaning up the burning of coal would be the best way to make growth greener

IS GROWTH bad for the environment? It is certainly fashionable in some quarters to argue that trade and capitalism are choking the planet to death. Yet it is also nonsense. As our survey of the environment this week explains, there is little evidence to back up such alarmism. On the contrary, there is reason to believe not only that growth can be compatible with greenery, but that it often bolsters it.

This is not, however, to say that there are no environmental problems to worry about. In particular, the needlessly dirty, unhealthy and inefficient way in which we use energy is the biggest source of environmental fouling. That is why it makes sense to start a slow shift away from today's filthy use of fossil fuels towards a cleaner, low-carbon future.

There are three reasons for calling for such an energy revolution. First, a switch to cleaner energy would make tackling other green concerns a lot easier. That is because dealing with many of these—treating chemical waste, recycling aluminium or incinerating municipal rubbish, for instance—is in itself an energy-intensive task. The second reason is climate change. The most sensible way for governments to tackle this genuine (but long-term) problem is to send a powerful signal that the world must move towards a low-carbon future. That will spur all sorts of innovations in clean energy.

The third reason is the most pressing of all: human health. In poor countries, where inefficient power stations, sooty coal boilers and bad ventilation are the norm, air pollution is one of the leading preventable causes of death. It affects some of the rich world too. From Athens to Beijing, the impact of fine particles released by the combustion of fossil fuels, and especially coal, is among today's biggest public-health concerns.

Dethroning King Coal

The dream of cleaner energy will never be realised as long as the balance is tilted toward dirty technologies. For a start, governments must scrap perverse subsidies that actually encourage the consumption of fossil fuels. Some of these, such as cash given by Spain and Germany to the coal industry, are blatantly wrong-headed. Others are less obvious, but no less damaging. A clause in America's Clean Air Act exempts old coal plants from complying with current emissions rules, so much of America's electricity is now produced by coal plants that are over 30 years old. Rather than closing this loophole, the Bush administration has announced measures that will give those dirty old clunkers a new lease on life. Nor are poor countries blameless: many subsidise electricity heavily in the name of helping poor people, but rich farmers and urban elites then get to guzzle cheap (mostly coal-fired) power.

That points to a second prescription: the rich world could usefully help poorer countries to switch to cleaner energy. A forthcoming study by the International Energy Agency estimates that there are 1.6 billion people in the world who are unable to use modern energy. They often walk many miles to fetch wood, or collect cow dung, to use as fuel. As the poor world grows richer in coming decades, and builds thousands of power plants, many more such unfortunates will get electricity. That good news will come with a snag. Unless the rich world intervenes, many of these plants will burn coal in a dirty way. The resultant surge in carbon emissions will cast a grim shadow over the coming decades. Ending subsidies for exporters of fossil-fuel power plants might help. But stronger action is probably needed, meaning that the rich world must be ready to pay for the poor to switch to low-carbon energy. This should not be regarded as mere charity, but rather as a form of insurance against global warming.

The final and most crucial step is to start pricing energy properly. At the moment, the harm done to human health and the environment from burning fossil fuels is not reflected in the price of those fuels, especially coal, in most countries. There is no perfect way to do this, but one good idea is for governments to impose a tax based on carbon emissions. Such a tax could be introduced gradually, with the revenues raised returned as reductions in, say, labour taxes. That would make absolutely clear that the time has come to stop burning dirty fuels such as coal, using today's technologies.

The dawning of the age of hydrogen

None of these changes need kill off coal altogether. Rather, they would provide a much-needed boost to the development of low-carbon technologies. Naturally, renewables such as solar and wind will get a boost. But so too would "sequestration", an innovative way of using fossil fuels without releasing carbon into the air.

This matters for two reasons. For a start, there is so much cheap coal, distributed all over the world, that poor countries are bound to burn it. The second reason is that sequestration offers a fine stepping-stone to squeaky clean hydrogen energy. Once the energy trapped in coal is unleashed and its carbon sequestered, energy-laden hydrogen can be used directly in fuel cells. These nifty inventions can power a laptop, car or home without any harmful emissions at all.

It will take time to get to this hydrogen age, but there are promising harbingers. Within a few years, nearly every big car maker plans to have fuel-cell cars on the road. Power plants using this technology are already trickling on to the market. Most big oil companies have active hydrogen and carbon-sequestration efforts under way. Even some green groups opposed to all things fossil say they are willing to accept sequestration as a bridge to a renewables-based hydrogen future.

Best of all, this approach offers even defenders of coal a realistic long-term plan for tackling climate change. Since he rejected the UN's Kyoto treaty on climate change, George Bush has been portrayed as a stooge for the energy industry. This week, California's legislature forged ahead by passing restrictions on emissions of greenhouse gases; a Senate committee has acted similarly. Mr Bush, who has made surprisingly positive comments about carbon sequestration and fuel cells, could silence the critics by following suit. By cracking down on carbon and embracing hydrogen, he could even lead.

Carbon sequestration

Fired up with ideas

Capturing and storing carbon dioxide could slow down climate change and also allow fossil fuels to be a bridge to a clean hydrogen-based future

IF THE world is to tackle the problem of climate change in earnest, "clean coal" has to become more than just an amusing oxymoron. All fossil fuels contain carbon, but coal is by far the most carbon-intensive. This is troubling, since global warming seems to be driven by an increase in the level of atmospheric greenhouse gases, of which carbon dioxide (CO_2) is the most worrisome. Coal is also the most abundant fossil fuel (see chart). If all known conventional oil and gas reserves (those in underground formations, obtained by drilling) were burned, the level of CO_2 in the atmosphere would still be less than twice what it was before the beginning of the industrial revolution. Climate change associated with that level of CO_2 might be tolerable. Burn all the coal, however, and it would be more than four times that starting-point—with larger, less predictable and quite likely more unpleasant climatic consequences.

Much of that coal will, nevertheless, be burned. In particular, poor countries such as China, India and South Africa have large reserves that are almost certain to be used to fuel economic growth. So it makes sense to consider possible technical fixes to the problem. These will never be the whole answer; unless the correct incentives are applied, burning coal without such fixes is likely to remain cheaper than burning it with them. But combined with the right incentives, in the form of such things as carbon taxes, they could help to keep the rise in atmospheric CO_2 to manageable proportions. They may also, surprisingly, help to usher in the green nirvana of a "hydrogen economy", in which the fuel of choice is that non-poisonous, non-greenhouse gas.

Catch me if you can

Technological solutions to rising atmospheric CO_2 (as opposed to, say, planting forests to soak the stuff up through photosynthesis) come in two parts. The first is capture: extracting the gas from the machine that is burning it. The second is sequestration: putting it somewhere it cannot easily escape from.

Capture is the more expensive of the two, especially when it needs to be designed into a plant from the start. One method that can be retrofitted on to existing machinery (although it is probably worth doing so only for large emitters) is to "scrub" CO_2 from an exhaust by passing it through a chemical, such as mono-ethanolamine, which has a particular affinity for the gas.

Smaller CO_2 sources (such as car engines and homes), which account for about half the gas generated by burning fossil fuels, are unlikely to be susceptible to retrofitted scrubbing. And even in large plants, scrubbing is easier and cheaper when the exhaust has a high concentration of CO_2—which is not the case for fuels that have been burned in air. Some have suggested using pure oxygen rather than air, but that is hardly an economically practical solution.

In some ways, though, a truly practical solution is even more radical: to change the way that energy is extracted from fossil fuels by separating the CO_2 formation from the process of heat generation. That can be done by adapting a well-established chemical process known as steam reformation. This involves reacting a carbon-based fuel with oxygen and steam to produce a so-called "synthesis gas" that is composed of carbon monoxide and hydrogen (much of the latter comes from the water, rather than the fuel). That mixture can be separated quite easily, and the hydrogen burned in, for example, a gas turbine. Then, mixing the carbon monoxide with more steam in the presence of a suitable catalyst yields CO_2 and still more hydrogen; again a mixture that can be separated quite easily.

It is this idea, known as the integrated gasifier combined cycle (IGCC) approach, that has environmentalists excited. It would require big changes to the design of energy-generating equipment, and so could not be introduced quickly. But it is not pie in the sky. And true visionaries will notice that, since it produces hydrogen, it permits the generation of electricity by fuel cells (chemical reactors that create electrical current from the reaction between hydrogen and oxygen, without any harmful emissions) as well as conventional gas turbines. Fuel cells are a critical component of most projections of what a hydrogen economy might look like.

Although the whole package seems idealistic, most of the technologies involved are in fact already in common use, in such processes as ammonia production. Indeed, there are several IGCC plants already operating in Europe and America (although they do not bother to remove CO_2). Some firms are talking of building one in China. Robert Williams, head of energy-systems analysis at Princeton University, reckons that, even in countries such as China, where tackling global warming is not exactly a priority, such gasification could lead to "zero emissions from coal", helped on its way by the extra revenues that may come from so-called "polygeneration" of clean synthetic fuels, hydrogen and electricity together.

Down under

Scrubbing and gasification can thus deliver CO_2 in a form that can be disposed of. Actually doing so, though, remains a challenge. Leaks from CO_2 repositories would hardly be as disastrous as leaks from a nuclear-waste dump. Nevertheless, to be effective, those repositories would have to stay gas-tight for centuries.

One way of disposing of the gas might be to **sink it beneath the waves**. The oceans already store a lot of CO_2, so they might be in-

Carbon in, Carbon out

Energy resources*	Resource base, GtC†	Potential CO_2 storage reservoirs: Sequestration option	Worldwide capacity‡ GtC†
Oil (conventional)	241		
Oil (unconventional, eg tar sands)	407	Ocean	1,000s
Natural gas (conventional)	253	Deep saline formations	100s-1,000s
Natural gas (unconventional, eg coal-bed methane)	509	Depleted oil and gas reservoirs	100s
Coal	5,151	Coal seams	10s-100s

*Proven reserves plus likely future resources †1 GtC = 1 billion tonnes of carbon equivalent ‡Orders of magnitude estimates

Sources: UNDP World Energy Assessment, 2000; Howard Herzog, Massachusetts Institute of Technology

duced to accept a little more. But it would have to be buried in water that is not likely to come to the surface any time soon. Some scientists worry, though, that dissolving vast quantities of CO_2 in the bottom of the ocean could result in ecological damage; others fear that the gas will be regurgitated wherever it is put. An international research consortium planned to test such worries by releasing 60 tonnes of CO_2 on the seabed near Hawaii—but noisy protests forced it to cancel the plan last month. Howard Herzog of the Massachusetts Institute of Technology, a member of the consortium, says that the team now hopes to shift to the North Sea.

A less speculative option would be to use **depleted oil and gas reservoirs**. These are layers of porous rocks topped by a cap of impermeable rock, usually in the shape of a dome. After decades of oil exploitation, there are plenty of ageing fields around. The advantages are many: the geology and technology involved are well understood; the exploration costs are small; and the reservoirs in question have already proved they can hold liquids and gases for aeons, since that is how the extracted hydrocarbons built up in them. What is more, injecting CO_2 into such wells could produce a saleable by-product; a similar technique is already used by oilmen to squeeze extra output from declining sources. For example, EnCana, a Canadian oil company, pays the Dakota Gasification Company to pump CO_2 produced at Dakota's coal gasification plant by pipeline to some of its wells.

In what may prove to be a straw in the wind, an American emissions brokerage called CO2e.com announced this week the largest-ever public trade in the emerging greenhouse-gas market. Ontario Power Generation, a Canadian company, bought the right to the emissions-reduction "credits" associated with 9m tonnes of CO_2. That gas, produced as a by-product of natural-gas processing, would normally have been vented into the atmosphere, but it will now be injected instead into old oil-fields in Wyoming, Texas and Mississippi. The power company has volunteered to cut its emissions of greenhouse gases, and sees these credits as a way to offset its fossil-fuel emissions.

A similar idea to burying CO_2 in old oil wells is to inject it into **coal seams** that are too deep and uneconomic to mine. This has two attractions. First, the injected gas will be absorbed on to the surface of the coal, and so locked up more or less permanently. Second, the incoming CO_2 often displaces methane that would otherwise not have seen the light of day. Capturing and selling that methane could turn this approach into a nice little earner. A scheme in New Mexico already uses CO_2 in this manner.

Also promising are **saline aquifers** located deep below the earth's surface. CO_2 pumped into such places would dissolve, at least in part, in the salt water. In some such formations, it would also react with local silicate minerals to form carbonates and bicarbonates that could stay put for millions of years. Statoil, a Norwegian oil firm, has been pumping CO_2 into a deep saline aquifer under the North Sea since 1996—the first time "geological" sequestration of this sort has been motivated by a fear of climate change, in the form of a Norwegian tax on carbon emissions. That tax created the incentive for Statoil to bury the stuff rather than continue releasing it into the atmosphere.

Geological sequestration, then, is not merely a speculative idea. In some cases it may even pay part or all of its own way, by releasing otherwise inaccessible fuel deposits. Recognising this, eight big energy companies, led by BP, have recently formed the CO_2 Capture Project to promote research. The Natural Resources Defence Council, a big American green group, says it is keeping an open mind about this sort of sequestration. And there is one other important endorsement: "We all believe technology offers great promise to significantly reduce emissions—especially carbon capture, storage and sequestration technologies." Thus spake George Bush, the man who said No to the Kyoto treaty on climate change.

Trading FOR Clean Water

States and localities are intrigued by proposals to create market mechanisms for solving intractable water-pollution problems.

By Tom Arrandale

For thousands of years, the shady forests that grew along the banks of Oregon's Tualatin River kept the water chilly enough for salmon, steelhead and cutthroat trout. The temperature began to rise, however, as pioneer farmers cleared dense tree stands to plant fields, and subsequent development channeled surface runoff into the 700-square-mile watershed. The Tualatin got warmer still when two sewage-treatment plants were built in the 1970s and started discharging 70-degree effluent into its waters. Now, federal and state regulators are holding Clean Water Services, the multi-county district that serves the suburbs west of Portland, responsible for cooling the Tualatin down so native cold-water fish can swim upstream for spawning.

One solution would be for CWS to spend $150 million on refrigeration equipment to chill the treated effluent before it is released. But the agency recently received permission to try a cheaper, and quite possibly more effective, alternative: It plans to start paying local farmers $1.5 million over the next five years to plant Douglas fir, red alder, Pacific willow and other native trees along the Tualatin and its upstream tributaries. For the nation's water-pollution police, "it's really radical letting somebody plant trees instead of doing technology-based things," says Sonja Biorn-Hansen, of the Oregon Department of Environmental Quality.

In return for allowing Clean Water Services to scrap plans for installing the costly equipment, Oregon water-quality officials will require the agency to bring temperatures all along the river down twice as much as chillers alone would have accomplished. "Our regulators gave us 20 years for the trees to grow," but to meet that goal "we'll need somewhere between 15 and 40 miles of shade," says Bruce Cordon, the district's project coordinator.

In addition to reducing Clean Water Service's costs by better than 90 percent, the arrangement will help curtail soil erosion, control polluted farm runoff, replenish streamside habitat and help salvage the state's imperiled salmon and trout populations. This tree-planting approach is one of the nation's boldest experiments in relying on old-fashioned market incentives to control pollution as efficiently as possible.

In the past decade, full-fledged exchanges have developed for trading air-emissions credits to address acid rain from Midwestern power plants and more recently for greenhouse gases. The Tualatin River deal and other similar examples are generating interest in a systematic approach to swapping effluent credits—at least within individual watersheds—as a way of solving intractable water-pollution problems.

SHARING THE CLEAN-UP BURDEN

Despite enormous investments in cleaning up factories, sewage-treatment plants and other end-of-pipe pollution sources, more than half of the nation's 2,000 watersheds still don't comply with federal Clean Water Act goals for making waters fit for fishing, swimming and drinking. That's largely because pollutants running off farmers' fields, golf courses, parks and lawns are still clogging streams and lakes with sediment, raising their temperatures, and clouding waters with excessive loads of phosphorus, nitrogen and other oxygen-consuming nutrients. Landowners who release these widely dispersed "non-point" pollutants are exempt from the 30-year-old law requiring corporations and municipal sewage systems to obtain government permits that force them to clean up the effluent they release.

In an effort to fill in the regulatory gap, pollution-control agencies have begun the process of enforcing "total maximum daily loads," which impose contaminant limits and temperature standards for streams, rivers and lakes. They've also begun broadening the regulatory focus to take in entire watersheds, including contaminated non-point runoff as well as point-source releases, and designing strategies for spreading the burden of bringing pollutant loads down to TMDL limits.

Local sewage agencies and businesses will still be on the regulatory hook if tighter limits go into effect. But ratcheting point-source discharges down even more to comply with tighter TMDLs would be an expensive proposition. At the same time, farmers and other landowners can reduce their own "non-point" runoff much more cheaply by adopting sensible soil-management practices. The U.S. Environmental Protection Agency calculates that the nation could save $900 million and still meet water-quality goals with trading mechanisms that let big polluters buy marketable credits from landowners who can then use the cash to curtail "non-point" runoff from their property.

EPA has been backing trading experiments since 1996. Minnesota approved a prototype trading venture in a 1997 permit that allowed Rahr Malting Co. to start discharging barley-processing byproducts into the lower Minnesota River. To offset the 150 pounds of nutrients it releases daily, the company has spent $255,000 to keep twice that much out of the river by paying four farms within the watershed to plant native grasses, build livestock fences and adopt other measures to curb erosion that would otherwise carry nutrients into the system.

For the most part, however, trading proposals to cut the costs of controlling water pollutants have yet to catch on. In an analysis published last year of 37 pilot programs, a consultant and a University of Maryland professor concluded that "despite the compelling economic logic . . . very few nutrient-credit trades have actually taken place."

Connecticut has implemented a limited pollutant-trading scheme among 79 sewage-treatment plants to help them deal with steadily tightening limits on nutrient-laden effluent that reaches Long Island Sound. The state's sewage agencies must collectively reduce their nitrogen discharges by nearly two-thirds to reverse nutrient loads that are depleting oxygen levels in the sound's waters. Trading began in 2002, with the state setting the value of credits that plants must acquire if they don't meet discharge standards as the limits are ratcheted down; the state in turn buys excess credits that other facilities earn because they have already invested in tighter controls and therefore are reducing nitrogen discharges below the current standard.

Advocates of market mechanisms think pollutant trading will have to play a much more significant role for governments to supplement point-source controls by finally curbing non-point releases as well. Michigan, Maryland and Virginia have launched pilot trading efforts for controlling farmland nutrients that wash into the Great Lakes and Chesapeake Bay. North Carolina set up a nutrient trading system in 1991 for 14 point-source dischargers in the Tar-Pamlico Basin; the program finances state assistance for farmers to control phosphorus and nitrogen runoff, but no trading has taken place.

Around Colorado's Lake Dillon reservoir, resort towns have adopted phosphorus controls and sold credits to sewage systems. Outside Denver, a few small-scale phosphorus trades have been made along the Cherry Creek Basin, but a full-fledged market hasn't developed. EPA and Idaho regulators are backing the Boise sewage system's proposal for phosphorus trading on the Lower Boise River with local farmers, but the plan was stymied several years while federal and state officials debated setting TMDL levels for phosphorus downstream in the Snake River system.

Wisconsin's Natural Resources Department has conducted studies to assess the potential for similar arrangements between cities and farmers along three major watersheds. So far, however, just one municipality, Cumberland (population 2,400), has taken the plunge into trading with nearby farmers. The projected cost of installing phosphorus controls at the city's 450,000-gallons-per-day sewage treatment plant was $150,000, plus $35,000 a year for chemicals and operating expenses. For just $20,000 a year, the city is instead paying farmers to cover the cost of shifting roughly 1,000 acres to no-till planting. That has kept twice as much phosphorus from eroding into the Red Cedar watershed as retrofitting the treatment works would have accomplished. Cumberland and Wisconsin DNR officials acknowledge that trading works only because Barron County's soil conservation manager has been doing all the legwork to identify potential trades at no cost to city coffers. Other Wisconsin cities found trading doesn't make economic sense for meeting the state's phosphorus limit, at least at its current level.

THE TROUBLE WITH TRADES

To function, any market requires both willing buyers and willing sellers. The Bush administration is pumping money into new pilots, but EPA and many states have yet to follow through on imposing rigid TMDLs that would stimulate demand for effluent credits from factories and sewage systems. Meanwhile, controlling non-point runoff remains voluntary; and direct federal and state subsidies for best management practices make trading less attractive for landowners. What's more, enforcing trades requires considerable work by state regulators to make sure goals are met. "I like pollution trading," says James Klang, a Minnesota Pollution Control Agency engineer who monitors the Rahr Malting agreement, "but there is upfront knowledge of the watershed that you have to have before you enter into it."

Environmentalists, meanwhile, contend that trading in effect grants big point sources a right to keep polluting more than they should while rewarding farmers for improvements they should be making anyway. "The way we see it, there's nothing to trade, because their responsibilities are already there," says Brian Wegener, the watershed coordinator for Tualatin Riverkeepers, an Oregon conservation organization. Despite such doubts, Minnesota is working to renew the Rahr permit; and the planned trading efforts seem likely to come together in the next year on both the Tualatin and Boise rivers in the Pacific Northwest.

"You're never going to see a big market like you have in air," says Claire Schary, a former EPA acid-rain trading expert who now oversees the Tualatin and Boise proposals from the agency's regional office in Seattle. But, she adds, trading still makes sense for dealing with some pollutants in watersheds "if you have a big point source that's bumping up against a limit in its permit."

EVERY STATE IS A COASTAL STATE

BY JENNY CARLESS

From the farmer in Iowa to the shrimper in Louisiana, everyone plays a role in the health of our oceans.

At certain times of the year, oxygen levels in the Gulf of Mexico fall so low that you'd be hard pressed to find shrimp or many fish species living in this aptly named "Dead Zone." Far away in the cold, remote reaches of Alaska—an isolated area prized for its pristine nature and beauty—contaminants like DDT, PCBs, and heavy metals are accumulating in coastal environments and creating serious health concerns for humans and wildlife alike.

What do these two seemingly distinct environmental concerns have in common? Their causes can come from hundreds—even thousands—of miles away. Nitrogen from far up the Mississippi River and its tributaries in the Midwest is the largest culprit in creating the Dead Zone, and pollution released into the air on distant continents is fouling the Arctic region.

These two cases prove that a state doesn't need a stretch of beach along its borders to qualify as a coastal state. Today, every state is a coastal state. Pollutants from household, garden, agricultural, municipal, and industrial use can travel downstream, across land with rainfall, and through the atmosphere to reach the ocean, no matter where they were originally released into the environment. "Individuals have a huge impact on the marine environment, but many don't realize it," says Tracy Kuhn, a Louisiana shrimper. "What each of us does may not seem much, but when you take what every household on the planet does, it adds up to a lot."

Ecosystems and *People Are Harmed*

Kuhn has been shrimping for many years and knows firsthand about the effects of the fertilizers and other chemicals that pollute the coastal zone. "When the Dead Zone is out there, we have to travel much farther to get outside it," she explains. "Our fuel expense increases, we're away longer, and we have to adjust our boats and equipment for the deeper water. Catching the same amount of shrimp costs a lot more."

The Dead Zone

Between the 1950s and the 1980s, the amount of nitrogen reaching the Gulf of Mexico tripled. Nitrogen, of course, is the main ingredient of fertilizer—it enriches soil and makes it more fertile. Nitrogen also enriches water. But excess nitrogen fuels a massive bloom of algae and phytoplankton. Those organisms, which usually produce oxygen, use up oxygen when decomposed by bacteria and other micro-organisms.

"The Gulf ecosystem produces more phytoplankton, driven by nitrogen loading from the Mississippi River drainage basin, and oxygen levels drop," explains Dr. Nancy Rabalais, a scientist at the Louisiana Universities Marine Consortium. The result is *hypoxia*, or very low oxygen. Most organisms require oxygen to survive, so an area that is hypoxic is truly a "dead zone." The Dead Zone in the Gulf varies in dimensions from month to month and year to year, but in the past, this seasonal hypoxic area has grown to as large as 20,000 square kilometers (12,000 square miles)—larger than the state of Massachusetts.

Fifty-six percent of the nitrogen entering the Mississippi River drainage basin begins its journey north of the confluence with the Ohio River—hundreds of miles away from the Gulf.

Hypoxia may have existed at some level before the 1940s and 1950s, but it has intensified since then. "The concentration of nitrogen has gone up in relation to the number of people and the amount of agriculture in the watershed," Rabalais explains. "There are other causes—including loss of wetlands and riparian (or riverbank) buffers, atmospheric deposition, and urban runoff—but over time, one of the biggest changes has been in the quantity of fertilizers used."

Fifty-six percent of the nitrogen entering the Mississippi River drainage basin begins its journey north of the confluence with the Ohio River—hundreds of miles away from the Gulf.

The Gulf of Mexico is not alone. According to research Rabalais and her colleagues have conducted, hypoxic and anoxic (with no oxygen) waters in shallow coastal and estuarine areas appear to be increasing, and now affect more than half of U.S. waters. Currently, the world's largest hypoxic zone—about 70,000 square kilometers—can be found in the Baltic basins, the body of water separating Sweden and Poland.

By Land and Air: How Urban Runoff and Air Pollution Affect the Oceans

Farmers should not be singled out for blame. Homeowners seeking lush green lawns can overfertilize, and these fertilizers, too, reach coastal waters. Pollution also travels through the air and over land, so there are many different sources of coastal contamination.

As urban development covers more land with impenetrable surfaces like roofs, pavement, and roads, storm water has less opportunity to be filtered by forests, plants, and soil before reaching waterways. Rainfall collects the contaminants that build up on these impervious surfaces during dry weather and flows, untreated, through storm drains into rivers and streams and then out to the ocean. Oil, grease, garden fertilizers and pesticides, industrial and household chemicals, pet droppings, and litter are just some of these contaminants.

Pollution traveling through the atmosphere is of equal concern. In general, atmospheric pollutants are caused by human activity, and our need for energy. Sources include refineries, power plants, cars, planes, and trains. Gaseous chemicals, in particular, are persistent and travel great distances.

The persistent organic pollutants (POPs) that have appeared in remote regions of Alaska are of great concern in that state, where many natives in coastal areas subsist on animals high on the food chain and/or high in fat, such as seals, walrus, salmon, and whales. "POPs and heavy metals become concentrated in living organisms, persist longer in our northern environment, and often accumulate in organs and fatty tissues," explains Tony Knowles, former Governor of Alaska and a Pew Oceans Commissioner. "Although most of these chemicals have been banned in the U.S., they travel long distances by air and water currents from all parts of the world and come to the Arctic, where they settle in our cold climate and break down very slowly."

Why Should I Care?

If every state is a coastal state, then every individual can and should work towards protecting and cleaning up the oceans and coastal zones.

"Why should I care about the ocean when I live in Ohio?" Dr. Kathryn Sullivan asks rhetorically. Also a Pew Oceans Commissioner, Sullivan's experience as both a former astronaut and the former chief scientist for the National Oceanic and Atmospheric Administration gives her a 'big picture' understanding of global interconnectedness. "The simple answer is because about half of the variability each year in our Ohio weather—in other words whether it's hot or cold, dry or wet—is driven by that sloshing water in the central Pacific Ocean known as El Niño." And the truth is, we humans affect the ocean as much as it affects us.

This awareness is driving many individuals, groups, and government agencies across the nation to address their role in coastal pollution. Landowners, for example, have a great opportunity to help reduce the impact of agricultural operations on the downstream environment. Since 1960, the amount of nitrogen in the environment has increased two to three times. This stunning increase is due to the creation of synthetic fertilizer and the burning of fossil fuels.

Chemical fertilizers typically use excessive amounts of inorganic nitrogen. Since they provide much more nitrogen than the plants actually need, ground and surface water is polluted with excess nitrates, the nitrogen-containing compounds in fertilizers. Nitrates have been proven to be harmful to humans and animals alike, and also contribute to water contamination.

Dan Specht, an organic farmer in McGregor, Iowa, thinks that organic farming can help address nitrogen runoff, among other benefits. "Organic rotations essentially require that the nitrogen come from natural, biological sources, which tend not to be converted instantaneously into nitrates," he explains. "They tend to be more stable, and a higher percentage of the nitrogen in the soil is from recently excreted nutrients from living matter. That's how nitrogen is naturally released, so it doesn't get converted into nitrates all at once." This means that organic farming isn't contributing to the nitrogen surplus.

Many agricultural practices can reduce nutrient runoff and losses to groundwater. For example, soil erosion (and resultant nitrogen loss) can be reduced by adjusting the timing of crop cultivation, changing drainage patterns, planting cover crops, adding wetlands, and implementing practices such as contour plowing, stream-bank protection, and grazing management.

Organic rotations require nitrogen from natural, biological sources, which tend not to be converted instantaneously into nitrates. It's a win-win-win situation. If farmers prevent the nitrogen loss, it helps their pocketbooks.

The benefits aren't just found downstream. "I get to witness an abundance of healthy wildlife on my farm," says Specht, "and I see the ground soak up water like a thirsty sponge, even in heavy downpours. Meanwhile, similarly sloped fields on other farms have brown gullies full of water pouring off."

Specht is optimistic about working towards a solution to nutrient loss. "It's a win-win-win situation," he says. "Farmers pay for the nitrogen they lose, so if they can prevent the loss, it helps their pocketbook. It helps Iowa water users who have to battle high nitrates in the water, and it's going to help folks downstream along the Gulf of Mexico, including the fishermen."

Restoring Riparian Areas

In Maryland, the Conservation Reserve Enhancement Program promotes establishing buffers, restoring wetlands, and retiring highly erodible agricultural land next to streams and other water bodies. Through this government program, landowners can receive annual cash payments in exchange for taking sensitive land out of agricultural production and for restoring and protecting these important buffer areas. Several other states offer similar programs.

Shedding Light on the Black Sea

Reducing nutrient loads can contribute to stunning recoveries, as evidenced in the Black Sea. Just like the Mississippi River, the northwest shelf of the Black Sea, where the Danube enters, had seen an increase of nutrient deposition since the 1950s. And at one time a large, low-oxygen zone existed in the Black Sea similar in size to the Gulf of Mexico's "Dead Zone." But now that hypoxic zone in the Black Sea is almost nonexistent.

What happened? "With the collapse of the Soviet Union, there were no more subsidies for fertilizer, so nitrogen and phosphorus levels in the Danube dropped by 50 percent," explains Rabalais. "That was a drastic cut in load. We're not proposing such severe cuts here, but it does show what can happen when you eliminate the fertilizer."

The Mississippi River/Gulf of Mexico Watershed Nutrient Task Force, a federal/state/tribal interagency group, is working towards integrating monitoring, modeling, and research in the areas affecting the Gulf. The group aims to reduce the size of the hypoxic zone and improve water quality within the Mississippi River basin.

Individuals Do *Make a Difference*

Success is possible in reducing urban runoff and nitrogen in the air, too. The Water Quality Protection Program (WQPP) for the Monterey Bay National Marine Sanctuary is a coalition of 25 federal, state, and local agencies and public and private groups. In its 10-year history, the WQPP has developed several initiatives to address urban runoff in the sanctuary's watersheds.

"One of the key things we've been involved in is public education—to help people make the connection between what they do in their back yard and driveway and the ocean," explains Dr. Holly Price, resource protection coordinator with the Sanctuary. "Another is getting volunteers out to sample water quality."

About half of the variability each year in Ohio weather is driven by that sloshing water in the central Pacific Ocean known as El Niño. And the truth is, we humans affect the ocean as much as it affects us.

The Monterey Bay Sanctuary Citizen Watershed Monitoring Network supports about 200 monitors who gather data throughout the year and at special events such as the 'First Flush'—in which volunteers collect urban storm water runoff from the first major rain of the season. First flush samples are monitored for a variety of items, including bacteria, nutrients, oil and grease, total dissolved solids, total suspended solids, zinc, copper, and lead. "Citizen water quality monitoring data are becoming a valuable resource for management decisions," says Bridget Hoover, the network's coordinator. "Local jurisdictions use the data to help identify problem areas and improve management decisions."

Citizens of an Ocean State

The good news is that, with a few exceptions, many effects of marine pollution are reversible. But making improvements requires an ethic by which people, no matter where they live, take responsibility for their part in causing pollution.

It's in everyone's interest to do so. As Kathryn Sullivan advises, "When half of the fundamental heartbeat where you live comes from the ocean, you'd be very wise to consider yourself a citizen of an ocean state."

WHAT YOU CAN DO

We can all help clean up urban runoff. Take motor oil, antifreeze, and other hazardous materials to proper collection sites and dispose of pet waste in the garbage can. Use household and garden chemicals and lawn fertilizer sparingly, and use a broom instead of a hose to clean up yard waste. Wash your car on the grass, or take it to a car wash; they are required to treat the water they use before releasing it.

You can help reduce atmospheric pollution, too. Make sure your next car gets high mileage, and consider biking, carpooling and using public transportation. Encourage and support legislation for renewable energy sources, better mileage requirements, and stricter emission control on vehicle and industrial discharges

UNIT 3

The Region

Unit Selections

15. **The Rise of India**, Manjeet Kripalani and Pete Engardio
16. **Between the Mountains**, Isabel Hilton
17. **A Dragon With Core Values**, The Economist
18. **L.A. Area Wonders Where to Grow**, John Ritter
19. **Reinventing a River**, Cait Murphy and Roseanne Haggerty
20. **Unscrambling the City**, Christopher Swope
21. **An Inner-City Renaissance**, Aaron Bernstein, Christopher Palmeri and Roger O. Crockett
22. **On the Road to Agricultural Self-Sufficiency**, Saudi Arabia

Key Points to Consider

- To what regions do you belong?
- Why are maps and atlases so important in discussing and studying regions?
- What major region in the world is experiencing change? Which ones seem not to change at all? What are some reasons for the differences?
- What regions in the world are experiencing tensions? What are the reasons behind these tensions? How can the tensions be eased?
- Can urban areas like Los Angeles continue to grow in population and area? What problems accompany growth?
- Why are regions in Africa suffering so greatly?
- Will agricultural output keep pace with population growth in the 21st century?
- Why is regional study important?

Links: www.dushkin.com/online/

These sites are annotated in the World Wide Web pages.

AS at UVA Yellow Pages: Regional Studies
http://xroads.virginia.edu/~YP/regional.html

Can Cities Save the Future?
http://www.huduser.org/publications/econdev/habitat/prep2.html

IISDnet
www.iisd.org

NewsPage
http://www.individual.com

Treaty on Urbanization
http://www.geocities.com/atlas/urb/tretyurb.html

Virtual Seminar in Global Political Economy/Global Cities & Social Movements
http://csf.colorado.edu/gpe/gpe95b/resources.html

World Regions & Nation States
http://www.worldcapitalforum.com/worregstat.html

The region is one of the most important concepts in geography. The term has special significance for the geographer, and it has been used as a kind of area classification system in the discipline.

Two of the regional types most used in geography are "uniform" and "nodal." A uniform region is one in which a distinct set of features is present. The distinctiveness of the combination of features marks the region as being different from others. These features include climate type, soil type, prominent languages, resource deposits, and virtually any other identifiable phenomenon having a spatial dimension.

The nodal region reflects the zone of influence of a city or other nodal place. Imagine a rural town in which a farm-implement service center is located. Now imagine lines drawn on a map linking this service center with every farm within the area that uses it. Finally, imagine a single line enclosing the entire area in which the individual farms are located. The enclosed area is defined as a nodal region. The nodal region implies interaction. Regions of this type are defined on the basis of banking linkages, newspaper circulation, and telephone traffic, among other things.

This unit presents examples of a number of regional themes. These selections can provide only a hint of the scope and diversity of the region in geography. There is no limit to the number of regions; there are as many as the researcher sets out to define.

The first article highlights "brainpower" in India and the country's rapid economic growth. "Between the Mountains" documents the struggle between India and Pakistan over Kashmir. The next article considers the competition between China and Hong Kong. Continued growth in Southern California is discussed in the next article. The next article discusses the renewal of the Merrimac River in New England. Positive growth in inner-city ghettos is discussed next. The final article reviews the success of the Saudi Arabian agricultural sector.

THE RISE OF INDIA

Growth is only just starting, but the country's brainpower is already reshaping Corporate America

By Manjeet Kripalani and Pete Engardio

PULLING INTO GENERAL ELECTRIC'S John F. Welch Technology Center, a uniformed guard waves you through an iron gate. Once inside, you leave the dusty, traffic-clogged streets of Bangalore and enter a leafy campus of low buildings that gleam in the sun. Bright hallways lined with plants and abstract art—"it encourages creativity," explains a manager—lead through laboratories where physicists, chemists, metallurgists, and computer engineers huddle over gurgling beakers, electron microscopes, and spectrophotometers. Except for the female engineers wearing saris and the soothing Hindi pop music wafting through the open-air dining pavilion, this could be GE's giant research-and-development facility in the upstate New York town of Niskayuna.

It's more like Niskayuna than you might think. The center's 1,800 engineers—a quarter of them have PhDs—are engaged in fundamental research for most of GE's 13 divisions. In one lab, they tweak the aerodynamic designs of turbine-engine blades. In another, they're scrutinizing the molecular structure of materials to be used in DVDs for short-term use in which the movie is automatically erased after a few days. In another, technicians have rigged up a working model of a GE plastics plant in Spain and devised a way to boost output there by 20%. Patents? Engineers here have filed for 95 in the U.S. since the center opened in 2000.

Pretty impressive for a place that just four years ago was a fallow plot of land. Even more impressive, the Bangalore operation has become vital to the future of one of America's biggest, most profitable companies. "The game here really isn't about saving costs but to speed innovation and generate growth for the company," explains Bolivian-born Managing Director Guillermo Wille, one of the center's few non-Indians.

The Welch center is at the vanguard of one of the biggest mind-melds in history. Plenty of Americans know of India's inexpensive software writers and have figured out that the nice clerk who booked their air ticket is in Delhi. But these are just superficial signs of India's capabilities. Quietly but with breathtaking speed, India and its millions of world-class engineering, business, and medical graduates are becoming enmeshed in America's New Economy in ways most of us barely imagine. "India has always had brilliant, educated people," says tech-trend forecaster Paul Saffo of the Institute for the Future in Menlo Park, Calif. "Now Indians are taking the lead in colonizing cyberspace."

> "Just like China drove down costs in manufacturing and Wal-Mart in retail, India will drive down costs in services," says an Indian IT exec

This techno take-off is wonderful for India—but terrifying for many Americans. In fact, India's emergence is fast turning into the latest Rorschach test on globalization. Many see India's digital workers as bearers of new prosperity to a deserving nation and vital partners of Corporate America. Others see them as shock troops in the final assault on good-paying jobs. Howard Rubin, executive vice-president of Meta Group Inc., a Stamford (Conn.) information-technology consultant, notes that big U.S. companies are shedding 500 to 2,000 IT staffers at a time. "These people won't get reabsorbed into the workforce until they get the right skills," he says. Even Indian execs see the problem. "What happened in manufacturing is happening in

WHERE **INDIA** IS MAKING AN IMPACT

SOFTWARE
India is now a major base for developing new applications for finance, digital appliances, and industrial plants.

IT CONSULTING
Companies such as Wipro, Infosys, and Tata are managing U.S. IT networks and re-engineering business processes.

CALL CENTERS
Thousands of Indians handle customer service and process insurance claims, loans, bookings, and credit-card bills.

CHIP DESIGN
Intel, Texas Instruments, and many U.S. startups use India as an R&D hub for mircroprocessors and multimedia chips.

...AND WHERE IT'S GOING **NEXT**

FINANCIAL ANALYSIS
Research for Wall street will surge as U.S. investment banks, brokerages, and accounting firms open big offices.

INDUSTRIAL ENGINEERING
India does vital R&D for GE Medical, GM, engine maker Cummins, Ford, and other manufactures plan big engineering hubs.

ANALYTICS
U.S. companies are hiring Indian math experts to devise models for risk analysis, consumer behavior, and industrial processes.

DRUG RESEARCH
As U.S. R&D costs soar, India is expected to be a center for biotechnology and clinical testing.

Data: *BusinessWeek*

services," says Azim H. Premji, chairman of IT supplier Wipro Ltd. "That raises a lot of social issues for the U.S."

No wonder India is at the center of a brewing storm in America, where politicians are starting to view offshore outsourcing as the root of the jobless recovery in tech and services. An outcry in Indiana recently prompted the state to cancel a $15 million IT contract with India's Tata Consulting. The telecom workers' union is up in arms, and Congress is probing whether the security of financial and medical records is at risk. As hiring explodes in India, the jobless rate among U.S. software engineers has more than doubled, to 4.6%, in three years. The rate is 6.7% for electrical engineers and 7.7% for network administrators. In all, the Bureau of Labor Statistics reports that 234,000 IT professionals are unemployed.

The biggest cause of job losses, of course, has been the U.S. economic downturn. Still, there's little denying that the offshore shift is a factor. By some estimates, there are more IT engineers in Bangalore (150,000) than in Silicon Valley (120,000). Meta figures at least one-third of new IT development work for big U.S. companies is done overseas, with India the biggest site. And India could start grabbing jobs from other sectors. A.T. Kearney Inc. predicts that 500,000 financial-services jobs will go offshore by 2008. Indiana notwithstanding, U.S. governments are increasingly using India to manage everything from accounting to their food-stamp programs. Even the U.S. Postal Service is taking work there. Auto engineering and drug research could be next.

WHY **CORPORATE AMERICA** IS BEATING A PATH TO INDIA

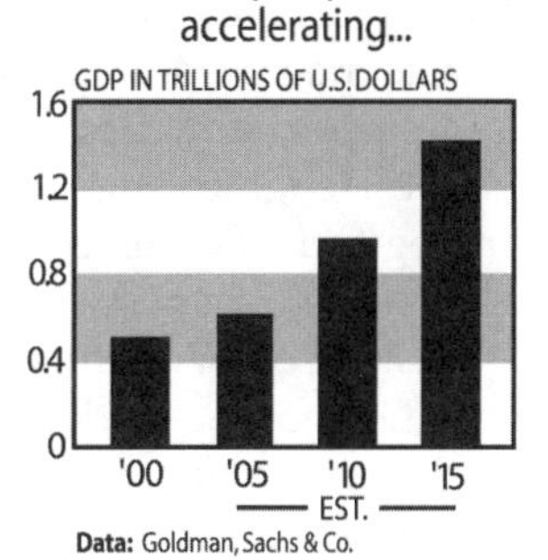

Data: Goldman, Sachs & Co.

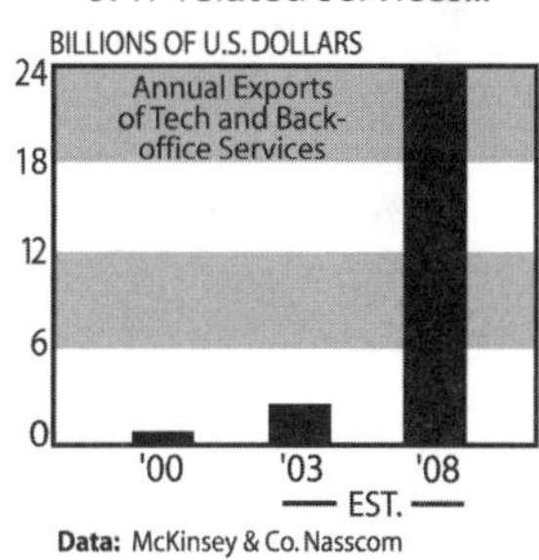

Data: McKinsey & Co. Nasscom

...a growing pool of tech graduates...

THOUSANDS

Engineers with College Degrees

640 / 480 / 320 / 160 / 0

'99 '01 '03 '05 '10 — EST. —

Data: Nasscom estimates

...and a swelling workforce

MILLIONS

Indians Aged 15-54

800 / 600 / 400 / 200 / 0

'01 '06 '13 — EST. —

Data: CLSA Emerging Markets, Statistical Outline of India

More Science in Schools

TECH LUMINARY Andrew S. Grove, CEO of Intel Corp., warns that "it's a very valid question" to ask whether America could eventually lose its overwhelming dominance in IT, just as it did in electronics manufacturing. Plunging global telecom costs, lower engineering wages abroad, and new interactive-design software are driving revolutionary change, Grove said at a software conference in October. "From a technical and productivity standpoint, the engineer sitting 6,000 miles away might as well be in the next cubicle and on the local area network." To maintain America's edge, he said, Washington and U.S. industry must double software productivity through more R&D investment and science education.

But there's also a far more positive view—that harnessing Indian brainpower will greatly boost American tech and services leadership by filling a big projected shortfall in skilled labor as baby boomers retire. That's especially possible with smarter U.S. policy. Companies from GE

WHO'S **BULKING** UP

Some of the biggest U.S. players in India

COMPANY	PURPOSE	INDIA STAFF
GE Captial Services	Back-office work	16,000
GE's John Welsh Tech Center	Product R&D	1,800
IBM Global Services	IT services, software	10,000
Oracle	Software, services	6,000**
EDS	IT services	3,500†
Texas Instruments	Chip design	900
Intel	Chip design, software	1,700
J.P. Morgan Chase	Back-office, analysis	1,200

*By 2005 **Unspecified †By 2004 Data: Company reports, Nasscom, Evalueserve

Medical Systems to Cummins to Microsoft to enterprise-software firm PeopleSoft that are hiring in India say they aren't laying off any U.S. engineers. Instead, by augmenting their U.S. R&D teams with the 260,000 engineers pumped out by Indian schools each year, they can afford to throw many more brains at a task and speed up product launches, develop more prototypes, and upgrade quality. A top electrical or chemical engineering grad from Indian Institutes of Technology (IITs) earns about $10,000 a year—roughly one-eighth of U.S. starting pay. Says Rajat Gupta, an IIT-Delhi grad and senior partner at consulting firm McKinsey & Co.: "Offshoring work will spur innovation, job creation, and dramatic increases in productivity that will be passed on to the consumer."

Whether you regard the trend as disruptive or beneficial, one thing is clear. Corporate America no longer feels it can afford to ignore India. "There's just no place left to squeeze" costs in the U.S., says Chris Disher, a Booz Allen Hamilton Inc. outsourcing specialist. "That's why every CEO is looking at India, and every board is asking about it." neoIT, a consultant advising U.S. clients on how to set up shop in India, says it has been deluged by big companies that have been slow to move offshore. "It is getting to a state where companies are literally desperate," says Bangalore-based neoIT managing partner Avinash Vashistha.

As a result of this shift, few aspects of U.S. business remain untouched. The hidden hands of skilled Indians are present in the interactive Web sites of companies such as Lehman Brothers and Boeing, display ads in your Yellow Pages, and the electronic circuitry powering your Apple Computer iPod. While Wall Street sleeps, Indian analysts digest the latest financial disclosures of U.S. companies and file reports in time for the next trading day. Indian staff troll the private medical and financial records of U.S. consumers to help determine if they are good risks for insurance policies, mortgages, or credit cards from American Express Co. and J.P. Morgan Chase & Co.

By 2008, forecasts McKinsey, IT services and back-office work in India will swell fivefold, to a $57 billion annual export industry employing 4 million people and accounting for 7% of India's gross domestic product. That growth is inspiring more of the best and brightest to stay home rather than migrate. "We work in world-class companies, we're growing, and it's exciting," says Anandraj Sengupta, 24, an IIT grad and young star at GE's Welch Centre, where he has filed for two patents. "The opportunities exist here in India."

If India can turn into a fast-growth economy, it will be the first developing nation that used its brainpower, not natural resources or the raw muscle of factory labor, as the catalyst. And this huge country desperately needs China-style growth. For all its R&D labs, India remains visibly Third World. IT service exports employ less than 1% of the workforce. Per-capita income is just $460, and 300 million Indians subsist on $1 a day or less. Lethargic courts can take 20 years to resolve contract disputes. And what pass for highways in Bombay are choked, crumbling roads lined with slums, garbage heaps, and homeless migrants sleeping on bare pavement. More than a third of India's 1 billion citizens are illiterate, and just 60% of homes have electricity. Most bureaucracies are bloated, corrupt, and dysfunctional. The government's 10% budget deficit is alarming. Tensions between Hindus and Muslims always seem poised to explode, and the risk of war with nuclear-armed Pakistan is ever-present.

So it's little wonder that, compared to China with its modern infrastructure and disciplined workforce, India is far behind in exports and as a magnet for foreign investment. While China began reforming in 1979, India only started to emerge from self-imposed economic isolation after a harrowing financial crisis in 1991. China has seen annual growth often exceeding 10%, far better than India's decade-long average of 6%.

In the Valley's Marrow

STILL, THIS DEEP SOURCE of low-cost, high-IQ, English-speaking brainpower may soon have a more far-reaching impact on the U.S. than China. Manufacturing—China's strength—accounts for just 14% of U.S. output and 11% of jobs. India's forte is services—which make up 60% of the U.S. economy and employ two-thirds of its workers. And Indian knowledge workers are making their way up the New Economy food chain, mastering tasks requiring analysis, marketing acumen, and creativity.

This means India is penetrating America's economic core. The 900 engineers at Texas Instruments Inc.'s Bangalore chip-design operation boast 225 patents. Intel Inc.'s Bangalore campus is leading worldwide research for the company's 32-bit microprocessors for servers and wireless chips. "These are corporate crown jewels," says Intel India President Ketan Sampat. India is even getting hard-wired into Silicon Valley. Venture capitalists say anywhere from one-third to three-quarters of the software, chip, and e-commerce startups they now back have Indian R&D teams from the get-go. "We can barely imagine investing in a company without at least asking what their plans are for India," says Sequoia Capital partner Michael Moritz, who nurtured Google, Flextronics, and Agile Software. "India has seeped into the marrow of the Valley [see box]."

It's seeping into the marrow of Main Street. This year, the tax returns of some 20,000 Americans were prepared by $500-a-month CPAs such as Sandhya Iyer, 24, in the Bombay office of Bangalore's MphasiS. After reading scanned seed and fertilizer invoices, soybean sales receipts, W2 forms, and investment records from a farmer in Kansas, Iyer fills in the farmer's 82-page return. "He needs to amortize these," she types next to an entry for new machinery and a barn. A U.S. CPA reviews and signs the finished return. Next year, up to 200,000 U.S. returns will be done in India, says CCH Inc. in Riverwoods, Ill., a supplier of accounting software. And it's not only Big Four firms that are outsourcing. "We are seeing lots of firms with 30 to 200 CPAs—even single practitioners," says CCH Sales Vice-President Mike Sabbatis.

A top electrical-engineering grad from one of the six Indian Institutes of Technology fetches about $10,000 a year

The gains in efficiency could be tremendous. Indeed, India is accelerating a sweeping reengineering of Corporate America. Companies are shifting bill payment, human resources, and other functions to new, paperless centers in India. To be sure, many corporations have run into myriad headaches, ranging from poor communications to inconsistent quality. Dell Inc. recently said it is moving computer support for corporate clients back to the U.S. Still, a raft of studies by Deloitte Research, Gartner, Booz Allen, and other consultants find that companies shifting work to India have cut costs by 40% to 60%. Companies can offer customer support and use pricey computer gear 24/7. U.S. banks can process mortgage applications in three hours rather than three days. Predicts Nandan M. Nilekani, managing director of Bangalore-based Infosys Technologies Ltd.: "Just like China drove down costs in manufacturing and Wal-Mart in retail," he says, "India will drive down costs in services."

GE Capital saves up to $340 million a year by performing some 700 tasks in India

But deflation will also mean plenty of short-term pain for U.S. companies and workers who never imagined they'd face foreign rivals. Consider America's $240 billion IT-services industry. Indian players led by Infosys, Tata, and Wipro got their big breaks during the Y2K scare, when U.S. outfits needed all the software help they could get. Indians still have less than 3% of the market. But by undercutting giants such as Accenture, IBM, and Electronic Data Systems by a third or more for software and consulting, they've altered the industry's pricing. "The Indian labor card is unbeatable," says Chief Technology Officer John Parkinson of consultant Cap Gemini Ernst & Young. "We don't know how to use technology to make up the difference."

Wrenching Change

MANY U.S. WHITE-COLLAR workers are also in for wrenching change. A study by McKinsey Global Institute, which believes offshore outsourcing is good, also notes that only 36% of Americans displaced in the previous two decades found jobs at the same or higher pay. The incomes of a quarter of them dropped 30% or more. Given the higher demands of employers, who want technicians adept at innovation and management, it could take years before today's IT workers land solidly on their feet.

India's IT workers, in contrast, sense an enormous opportunity. The country has long possessed some basics of a strong market-driven economy: private corporations, democratic government, Western accounting standards, an active stock market, widespread English use, and schools strong in computer science and math. But its bureaucracy suffocated industry with onerous controls and taxes, and the best scientific and business minds went to the U.S., where the 1.8 million Indian expatriates rank among the most successful immigrant groups.

Now, many talented Indians feel a sense of optimism India hasn't experienced in decades. "IT is driving India's boom, and we in the younger generation can really deliver the country from poverty," says Rhythm Tyagi, 22, a master's degree student at the new Indian Institute of Information Technology in Bangalore. The campus is completely wired for Wi-Fi and boasts classrooms with videoconferencing to beam sessions to 300 other colleges.

That confidence is finally spurring the government to tackle many of the problems that have plagued India for so long. Since 2001, Delhi has been furiously building a network of high-ways. Modern airports are next. Deregulation of the power sector should lead to new capacity. Free education for girls to age 14 is a national priority. "One by one, the government is solving the bottlenecks," says Deepak Parekh, a financier who heads the quasi-governmental Infrastructure Development Finance Co.

Future Vision

INDIA ALSO IS WORKING to assure that it will be able to meet future demand for knowledge workers at home and abroad. India produces 3.1 million college graduates a year, but that's expected to double by 2010. The number of engineering colleges is slated to grow 50%, to nearly 1,600, in four years. Of course, not all are good enough to produce the world-class grads of elite schools like the IITs, which accepted just 3,500 of 178,000 applicants last year. So there's a growing movement to boost faculty salaries and reach more students nationwide through broadcasts. India's rich diaspora population is chipping in, too. Prominent Indian Americans helped found the new Indian School of Business, a tie-up with Wharton School and Northwestern University's Kellogg Graduate School of Management that lured most of its faculty from the U.S. Meanwhile, the six IIT campuses are tapping alumni for donations and research links with Stanford, Purdue, and other top science universities. "Our mission is to become one of the leading science institutions in the world," says director Ashok Mishra of IIT-Bombay, which has raised $16 million from alumni in the past five years.

If India manages growth well, its huge population could prove an asset. By 2020, 47% of Indians will be between 15 and 59, compared with 35% now. The working-age populations of the U.S. and China are projected to shrink. So India is destined to have the world's largest population of workers and consumers. That's a big reason why Goldman, Sachs & Co. thinks India will be able to sustain 7.5% annual growth after 2005.

Skeptics fear U.S. companies are going too far, too fast in linking up with this giant. But having watched the success of the likes of GE Capital International Services, many execs feel they have no choice. Inside GECIS' Bangalore center—one of four in India—Gauri Puri, a 28-year-old dentist, is studying an insurance claim for a root-canal operation to see if it's covered in a certain U.S. patient's dental plan. Two floors above, members of a 550-strong analytics team are immersed in spreadsheets filled with a boggling array of data as they devise statistical models to help GE sales staff understand the needs, strengths, and weaknesses of customers and rivals. Other staff prepare data for GE annual reports, write enterprise resource-planning software, and process $35 billion worth of global invoices. Says GE Capital India President Pramod Bhasin: "We are mission-critical to GE." The 700 business processes done in India save the company $340 million a year, he says.

Indian finance whizzes are a godsend to Wall Street, too, where brokerages are under pressure to produce more independent research. Many are turning to outfits such as OfficeTiger in the southern city of Madras. The company employs 1,200 people who write research reports and do financial analysis for eight Wall Street firms. Morgan Stanley, J.P. Morgan, Goldman Sachs, and other big investment banks are hiring their own armies of analysts and back-office staff. Many are piling into Mindspace, a sparkling new 140-acre city-within-a-city abutting Bombay's urban squalor. Some 3 million square feet are already leased to Western finance firms. By yearend, Morgan Stanley will fill several floors of a new building.

For Silicon Valley startups, Indian engineers let them stretch R&D budgets. PortalPlayer Inc., a Santa Clara (Calif.) maker of multimedia chips and embedded software for portable devices such as music players, has hired 100 engineers in India and the U.S. who update each other daily at 9 a.m. and 10 p.m. J.A. Chowdary, CEO of PortalPlayer's Hyderabad subsidiary Pinexe, says the company has shaved up to six months off the development cycle—and cut R&D costs by 40%. Impressed, venture capitalists have pumped $82 million into PortalPlayer.

WHERE **CHINA** IS WAY AHEAD...	...WHERE **INDIA** HAS THE EDGE
GROWTH GDP has risen an average of 8% for the past decade, compared with India's 6%.	**LANGUAGE** English gives India a big edge in IT services and back-office work.
INFRASTRUCTURE Highways, ports, power sector, and industrial parks are far superior.	**CAPITAL MARKETS** Private firms have readier access to funding. China favors state sector.
FOREIGN INVESTMENT China lures $50 billion-plus a year. India gets $4 billion.	**LEGAL SYSTEM** Contract law and copyright protection are more developed than in China.
EXPORTS $266 billion reported in 2002 was more than four times India's total.	**DEMOGRAPHICS** Some 53% of India's population is under age 25, vs. 45% in China.

More Bang for the Buck

OLD ECONOMY COMPANIES are benefiting, too. Engine maker Cummins plans to use its new R&D center in Pune to develop the sophisticated computer models needed to design upgrades and prototypes electronically. Says International Vice-President Steven M. Chapman: "We'll be able to introduce five or six new engines a year instead of two" on the same $250 million R&D budget—without a single U.S. layoff.

The nagging fear in the U.S., though, is that such assurances will ring hollow over time. In other industries, the shift of low-cost production work to East Asia was followed by engineering. Now, South Korea and Taiwan are global leaders in notebook PCs, wireless phones, memory chips, and digital displays. As companies rely more on IT engineers in India and elsewhere, the argument goes, the U.S. could cede control of other core technologies. "If we continue to offshore high-skilled professional jobs, the U.S. risks surrendering its leading role in innovation," warns John W. Steadman, incoming U.S. president of Institute of

BRAINPOWER

India and Silicon Valley: Now the R&D Flows Both Ways

The ravages of the dot-com bust are still evident at Andale Inc.'s Mountain View (Calif.) headquarters. Half the office space sits abandoned, one corner of it heaped with discarded cubicle dividers and file cabinets. But looks are deceptive. The four-year-old startup, which offers software and research tools for online auction buyers and sellers, has seen its workforce nearly quadruple in the past year—with most of those jobs in Bangalore.

Andale's 155 workers in India, where employing a top software programmer runs a small fraction of the cost in the U.S., have been the key to the company's survival, says Chief Executive Munjal Shah, who grew up in Silicon Valley. In fact, Indian talent is adding vitality throughout Silicon Valley, where it's getting hard to find an info-tech startup that doesn't have some research and development in such places as Bangalore, Bombay, or Hyderabad. Says Shah: "The next trillion dollars of wealth will come from companies that straddle the U.S. and India."

The chief architects of this rising business model are the 30,000-odd Indian IT professionals who live and work in the Valley. Indian engineers have become fixtures in the labs of America's top chip and software companies. Indian émigrés have also excelled as managers, entrepreneurs, and venture capitalists. As of 2000, Indians were among the founders or top execs of at least 972 companies, says AnnaLee Saxenian, who studies immigrant business networks at the University of California at Berkeley.

Until recently, that brainpower mostly went in one direction, benefiting the Valley more than India. Now, this ambitious diaspora is generating a flurry of chip, software, and e-commerce startups in both nations, mobilizing billions in venture capital. The economics are so compelling that some venture capitalists demand Indian R&D be included in business plans from Day One. Says Robin Vasan, a partner at Mayfield in Menlo Park: "This is the way they need to do business."

The phenomenon is due in no small part to the professional and social networks Indians have set up in the Valley, such as The Indus Entrepreneurs (TiE), in Santa Clara: It now has 42 chapters in nine countries. Prominent Indians such as TiE founder and serial entrepreneur Kanwal Rekhi, venture capitalist Vinod Khosla, entrepreneur Kanwal Rekhi, and former Intel corp. executive Vin Dham serve as startup mentors and angel investors. In early November, Bombay-born Ash Lilani, senior vice-president at Silicon Valley Bank, led 20 Valley VCs on their first trip to India to scout opportunities. Of the bank's 5,000 Valley clients, 10% have some development work in India, but that's expected to rise to 25% in two years.

Such opportunities for the Valley's Indians flow both ways. Hundreds have returned to India since 2000 to start businesses or help expand R&D labs for the likes of Oracle, Cisco Systems, and Intel. The downturn—and Washington's decision to issue fewer temporary work visas—accelerated the trend. At a Nov. 6 tech job fair in Santa Clara, hundreds of engineers lined up, résumés in hand, for Indian openings offered by companies from Microsoft Corp. to Juniper Networks Inc. "The real development and design jobs are in India," says Indian-born job-seeker Jay Venkat, 24, a University of Alabama electrical engineering grad.

The deeper, more symbiotic relationship developing between the Valley and India goes far beyond the "body shopping" of the 1990s, when U.S. companies mainly wanted low-wage software-code writers. Now the brain drain from India is turning into what Saxenian calls "brain circulation," nourishing the tech scenes in both nations.

Some Valley companies even credit India with saving them from oblivion. Web-hosting software outfit Ensim Corp. in Sunnyvale relied on its 100-engineer team in Bangalore to keep designing lower-cost new products right through the downturn. "This company would not survive a day if not for the operation in India," says CEO Kanwal Rekhi. Before long, India may prove as crucial to the Valley's success as silicon itself.

—By Robert D. Hof in Santa Clara, Calif., with Manjeet Kripalani in Bombay

Electrical & Electronics Engineers Inc. That could also happen if many foreigners—who account for 60% of U.S. science grads and who have been key to U.S. tech success—no longer go to America to launch their best ideas.

Information-technology services could soon account for 7% of India's GDP

Throughout U.S. history, workers have been pushed off farms, textile mills, and steel plants. In the end, the workforce has managed to move up to better-paying, higher-quality jobs. That could well happen again. There will still be a crying need for U.S. engineers, for example. But what's called for are engineers who can work closely with customers, manage research teams, and creatively improve business processes. Displaced technicians who lack such skills will need retraining; those entering school will need broader educations.

Adapting to the India effect will be traumatic, but there's no sign Corporate America is turning back. Yet the India challenge also presents an enormous opportunity for the U.S. If America can handle the transition right, the end result could be a brain gain that accelerates productivity and innovation. India and the U.S., nations that barely interacted 15 years ago, could turn out to be the ideal economic partners for the new century.

—With Steve Hamm in New York

LETTER FROM KASHMIR

BETWEEN THE MOUNTAINS

India and Pakistan are caught in a dangerous struggle over Kashmir. But what do its people want?

BY ISABEL HILTON

When the French doctor François Bernier entered the Kashmir Valley for the first time, in 1665, he was astounded by what he found. "In truth," he wrote, it "surpasses in beauty all that my warm imagination had anticipated. It is not indeed without reason that the Moghuls call Kachemire the terrestrial paradise of the Indies." The valley, which is some ninety miles long and twenty miles across, is sumptuously fertile. Along its floor, there are walnut and almond trees, orchards of apricots and apples, vineyards, rice paddies, hemp and saffron fields. There are woods on the lower slopes of the surrounding mountains—sycamore, oak, pine, and cedar. The southern side is bounded by the Pir Panjal, not the highest mountain range in Asia but one of the most striking, rising abruptly from the valley floor. The northern boundary is formed by the Great Himalayas. At the heart of the valley lie Dal Lake and the graceful capital, Srinagar.

For Europeans, Kashmir became a locus of romantic dreams, inspiring writers like the Irish poet Thomas Moore, who didn't even need to visit it to understand its charms. "Who has not heard of the Vale of Cashmere," he wrote in 1817, "with its roses the brightest that earth ever gave." So seductive was this landlocked valley that, like a beautiful woman surrounded by jealous lovers, Kashmir attracted a succession of invaders, each eager to possess her.

Srinagar is a city of waterways, floating gardens, and lotus beds, spanned by nine graceful bridges.

The Moghuls established their control in the sixteenth century. Kashmir became the northern limit of their Indian empire as well as their pleasure ground, a place to wait out the summer heat of the plains. They built gardens in Srinagar, along the shores of Dal Lake, with cool and elegantly proportioned terraces—with fountains and roses and jasmine and rows of chinar trees. The Moghul rulers were followed by the Afghans and, later, by the Sikhs from the Punjab, who were driven out in the nineteenth century by the British, who then sold the valley, to the abiding shame of its residents, for seven and a half million rupees to the maharaja, Gulab Singh. Singh was the notoriously brutal Hindu ruler of Jammu, the region that lay to the south, beyond the Pir Panjal, on the edge of the plains of the Punjab.

Under Singh, the Kashmir Valley was conjoined in the princely state of Jammu and Kashmir. According to one calculation of the purchase, the ruler of the newly formed state had bought the people of Kashmir for approximately three rupees each, a sum he was to recover many times over through taxation. For the maharaja and his descendants and their visitors, the valley was luxurious paradise; they enjoyed fishing and duck shooting, boating excursions on Dal Lake, picnics in the hills and the saffron fields, moonlit parties in the magnificent gardens. In the penetrating cold of the winters, the visitors, and the maharaja, left the valley to itself and returned to Jammu.

Kashmir was also a natural crossroads. The Silk Route, with its great camel trains from China, passed to the north, and the country's mountain passes opened routes to the Punjab, Afghanistan, and Jammu. Through them successive intruders brought different cultures that added layers to Kashmir's own. The Kashmiri language was a mixture of Persian, Sanskrit, and Punjab; the handicrafts for which the valley was celebrated were Central Asian; and the religious faith was variously Buddhist, Hindu, Sikh, and Muslim. Sufi masters left a legacy of music and tolerance in their Muslim teachings. A Sikh who had lived many years in Srinagar described the culture of the valley as an old cloth so covered in patches that you can't see the original.

Today, the valley is predominantly Muslim, but, as part of the maharaja's portmanteau state of Jammu and Kashmir, it still shares its destiny with other faiths and peoples: the Hindus of Jammu, the Buddhists of Ladakh, as well as Gilgits and Baltis, Hunzas and Mirpuris. There had been conflicts between the communities in the past, but by the mid-twentieth century Kashmir was an unusually tolerant culture. It escaped the intercommunal violence that Partition brought to the neighboring Punjab when the British left the subcontinent, in 1947. Kashmir's violence was to occur later, as the two new states of India and Pakistan became the latest of Kashmir's neighbors to fight over it.

Today, Kashmir is partitioned—Pakistan controls slightly less than a third, India some sixty per cent, and China the rest. Most of Kashmir's twelve million people are concentrated in Indian-held territories, and the rest are mainly in Pakistan-held ones; relations among its many communities are now marked by mutual mistrust. And since the late eighties a bewildering number of combatants have fought a savage, irregular war that, in a steady daily toll of killing, has cost, depending on whom you believe, between thirty to eighty thousand lives. On the side of the Indian state, the participants include the local police, the Border Security Force, the Central Reserve Police Force, and the Army, supported by various intelligence organizations and a motley group of turncoat former militants who have muddied the public understanding of who, over the years, has done what to whom. Opposing them are a proliferation of Islamic militant groups. At one time, there were more than sixty of them. Several are fundamentalist and deadly—like the Lashkar-e-Taiba and Jaish-e-Mohammed, which are based in Pakistan (and have been listed as terrorists by the United States) and were recently banned by Pakistan's President Pervez Musharraf. The largest group, the Hizbul Mujahideen, is Muslim but not, its supporters insist, fundamentalist, and most of its activists, who number around a thousand, are Kashmiris.

Surrounding the insurgency is the wider, implacable hostility between India and Pakistan. But at its core is the story of a people who, for five centuries, have been longing to call their homeland their own.

Last October, I was permitted to go into what Pakistan calls Azad ("Free") Kashmir, a territory that Pakistan maintains is truly autonomous but which depends entirely on the country's military and money for its continued existence. India calls the territory Pakistan-occupied Kashmir. The entity has existed ever since Pakistan wrested this northwest third of the original state of Jammu and Kashmir from Indian control in a war that followed the 1947 Partition. For Pakistan, that war was the first step toward a liberation of Kashmir's Muslims from India. Once liberated, Pakistan hoped, the Kashmiris would join Muslim Pakistan.

At the time of Partition, Jammu and Kashmir was still ruled by a Hindu maharaja, Hari Singh, a descendant of Gulab Singh. The maharaja was one of five hundred and sixty-two fabulously rich feudal monarchs whom the British had manipulated in order to maintain their grip on much of India. At Partition, these states were given a choice of joining India or Pakistan. Independence was not on offer. Most joined India. The maharaja dithered for months, unable to decide between two equally unattractive options. As a Hindu, he did not like Pakistan. As an Indian, he did not like the British. As a prince, he cared neither for the antifeudal Mahatma Gandhi nor for the local Muslim leader, Sheikh Abdullah, who favored autonomy for Kashmir but without its maharaja. Then, on October 20, 1947, armed tribesmen and regular troops from Pakistan invaded Kashmir. The maharaja appealed to India for support and hastily agreed to sign the now famous Instrument of Accession to India: the state of Kashmir and Jammu was accepted as part of the new federal union of India; in exchange, it was, exceptionally, granted a semiautonomous status. (India would control only matters of defense, foreign affairs, and communications; everything else was to be run by Jammu and Kashmir's own parliament.) Pakistan, furious, refused to accept the legality of the accession, and Pakistan and India fought their first war over Kashmir.

In Pakistan, what is remembered was a promise made by the Indian Prime Minister Jawaharlal Nehru to hold a plebiscite in which the people of Kashmir could make their preferences clear. That plebiscite was never held. India blames Pakistan: in 1949, after a ceasefire was agreed to under United Nations supervision,

Pakistan failed to withdraw from Azad Kashmir, a betrayal that, India says, vitiated the commitment to the plebiscite.

Today, there are few routes that connect Azad Kashmir with Pakistan proper. Some fellow-journalists and I set out from Islamabad at 6:30 A.M. and drove for five hours along vertiginous valleys, through Muzaffarabad, the capital of Azad Kashmir, and on into the mountains to Chakothi, a town on what is now known as the Line of Control—the ceasefire line established in 1949, after that first war over Kashmir. There, we walked to a peaceful clearing and sat sipping fruit juice. An immaculately turned-out brigadier, Mohammed Yaqub, the commander of the sector, briefed us on the Pakistani version of the history of the present conflict.

Tensions were unusually high. The United States bombing of Afghanistan had begun, and the military's view was that India might take advantage of the situation—troop movements had been detected. Yaqub's list of the casualties incurred in the last thirteen years of what he saw as Kashmir's freedom struggle against India was startling, even if undoubtedly exaggerated: 74,625 killed, 80,317 wounded, 492 adults burned alive, 875 schoolchildren burned alive, 15,812 raped, 6,572 sexually incapacitated, 37,030 disabled, 96,752 missing.

We took a path that led to a bluff overlooking a tributary of the Jhelum River. There was a slender, deserted bridge. On the other side were the Indian Army fortifications. A line of washing flapped in a light breeze above a series of bunkers. I peered through binoculars at men peering through binoculars at me. They waved, I waved back. A Pakistani officer admitted that, in more relaxed times, he met his Indian counterparts on the bridge and shared tea and sweets. "We don't talk about the war," he said.

Just as night was falling, we stopped at a refugee camp about an hour's drive away. A camp manager called on the refugees to tell stories of the atrocities that had forced them from their homes in Indian-controlled Kashmir. The misery, no doubt, was real, but the exercise smelled too much of propaganda to be of any genuine interest. The message, though, was clear: Kashmir was the unspoken subtext of the Afghan war. Under President Musharraf, Pakistan has sided with the United States and backed the bombing of Afghanistan. Nearly twenty years earlier, Pakistan had also sided with the United States in its mission to end the Soviet occupation of Afghanistan, by enlisting Pakistan's Inter-Services Intelligence, the I.S.I., to arm and train Islamic warriors to lead the fight. The I.S.I. had seen the opportunity to foment discontent in Kashmir, and Islamic warriors were armed and trained and sent there as well. Both wars were seen as religious and patriotic causes. But now Musharraf had renounced the Taliban and his country's earlier ambition to dominate Afghanistan through support of its hard-line Islamist government. Would he also be forced to abandon a dream that Pakistan has clung to since 1947—of uniting the Muslims of Kashmir with the state of Pakistan?

I met a member of one of the Kashmiri militant groups in Islamabad. He called himself Iqbal, though we both knew that it was not his name. He was a good-looking man in his early forties, with black hair beginning to gray. We had arranged to meet in an outdoor café. He was nervous, and constantly scanned the customers until he insisted that we move to a different location. We drove around the city looking for somewhere to talk. Eventually, he took me to a house in an affluent district of the city, a two-story villa set back from the street by high walls. There, we sat on the floor, and he told me his story.

Iqbal had grown up in a Kashmir that preserved the memory—from before the Moghuls—of an independent country. For him, the Instrument of Accession was important because, in granting special autonomy, it implicitly acknowledged the idea of Kashmiri independence. But the Indian government, anxious about Pakistan's ambitions and uncertain of Kashmiri loyalty, regularly encroached on that autonomy. In 1953, Kashmir's popular Prime Minister, Sheikh Abdullah, was removed and arrested (he was suspected of autonomous leanings)—the first in a series of detentions that continued through the sixties. In 1963, a sacred relic—a hair of the Prophet's beard—disappeared from the Hazratbal mosque in Srinagar, and demonstrations erupted. The following year, India passed an order that allowed the Indian President to rule directly in Kashmiri affairs. By then, Muslim sentiments in the valley were hardening.

The long-established Kashmiri tradition of tolerance—the pluralism that had accommodated so many different faiths and cultures—was breaking down in the frustration generated by India's interference. One friend described to me what the valley was like in the late seventies and early eighties. There were, he recalled, fevered political discussions, stimulated by activist teachers who distributed everything from the works of the Egyptian Muslim Brotherhood to the teachings of Mao Zedong. Kashmiris were impatient for change. Their hopes were focussed on elections that were to take place in 1987.

The chief minister was Farooq Abdullah, who had returned to power after having been dismissed by Indira Gandhi, in 1984. He had regained his position by allying himself with the Indian National Congress Party, and many regarded him as a traitor to the cause of Kashmir. The opposition was led by the

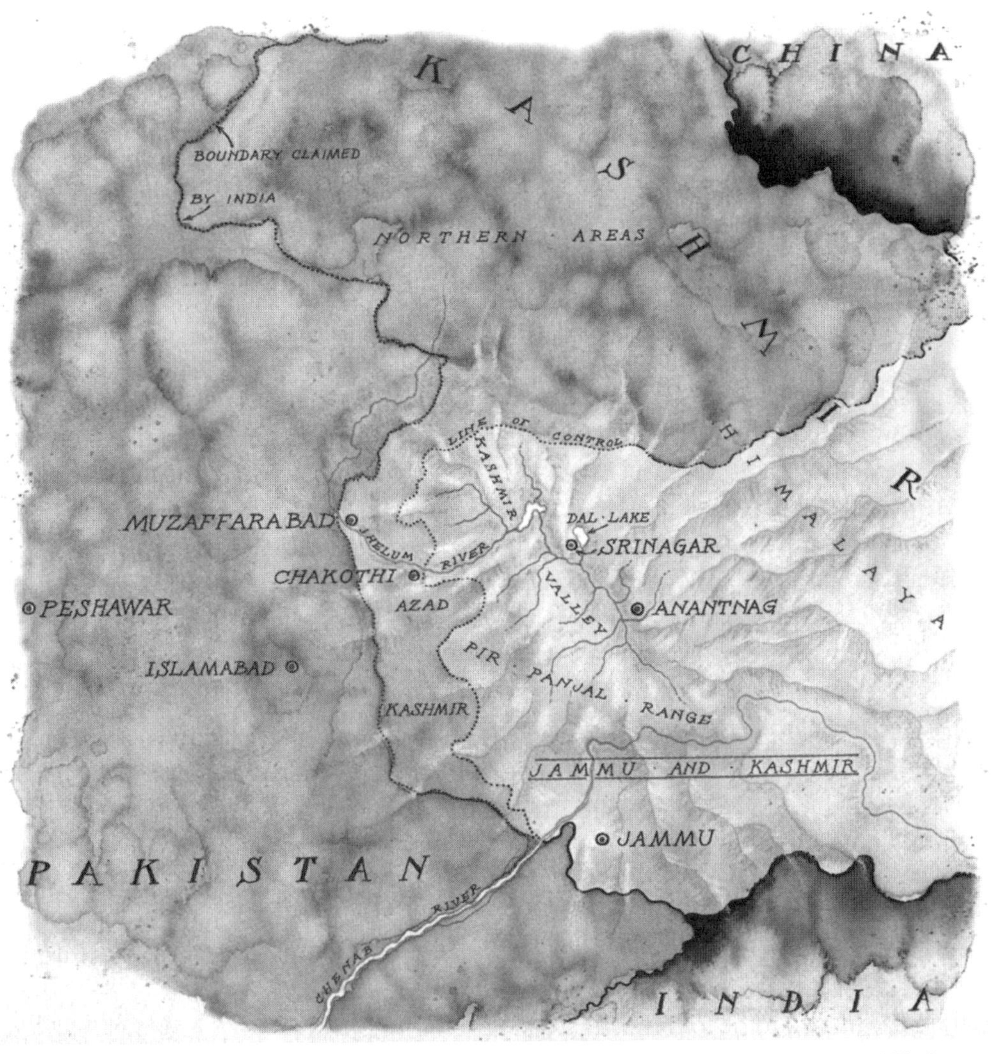

Mike Reagan

Kashmir has been occupied by Moghuls, Afghans, Sikhs, and the British, who sold it.

Muslim United Front, a coalition of ten or so Muslim parties campaigning on a platform of Islamic unity and greater autonomy for Kashmir.

Iqbal was a student at the university in Srinagar and was involved in the electoral campaign. "The 1987 elections were our last hope," he told me. Eighty per cent of the population of the valley turned out to vote. When the results were declared, Farooq Abdullah's pro-India Conference-Congress alliance had—to the dismay and disbelief of the voters—won a two-thirds majority.

The fraud had been crude and blatant. In one constituency in Srinagar, witnesses told me, the result had been publicly declared, only to be reversed an hour later. After the election, opposition candidates and party members were arrested. There were widespread street protests, which were brutally suppressed. "When the results were declared," Iqbal said, "people decided that we could not free our land through peaceful means."

Iqbal joined an underground group and was arrested. "I was in jail for two and a half years without trial," he said. When he was released, he was immediately rearrested and held for another two years. He was released and arrested again.

Iqbal returned to the university in Srinagar, but during an Army search an informer identified him as a militant. He was detained again. This time, he said, he was tortured. (According to Amnesty International, in Jammu and Kashmir torture has become so routine in the arrest-and-detention process that it is rarely reported.) But this time, once he was released, Iqbal took up arms. He joined Hizbul Mujahideen and spent three years as an underground militant. He was arrested three more times, before he finally escaped to Pakistan. He had spent, he said, fourteen years in prison, about half his adult life. "If you want to talk about Kashmir," he said, "you must talk about the eighty thousand innocent martyrs. It's a death rate of fifteen innocent civilians to every one Indian soldier."

A few weeks after I met with Iqbal, that balance shifted, marginally but dramatically, in the other direction. At eleven-forty on the morning of December 13th, in New Delhi, five men dressed in olive-green fatigues and armed with assault rifles, grenades, and explosives drove a white official car, complete with flashing lights and security passes, through the gates of the Indian Parliament complex. The session had just ended, and the politicians were beginning to disperse. It was only after security guards noticed the car turn the wrong way that they became suspicious. A guard ran after it, calling to the driver to stop. Alarmed, the Vice-President's security guard, waiting by his official vehicle, challenged the white car. Gunfire came from it as the car rammed into the Vice-President's vehicle, and the men inside ran toward the Parliament building. In the ensuing firefight, all five terrorists were killed, along with eight security personnel and a gardener. The car was found to be packed with explosives. The target of the assault was the Parliament building itself. Although the identity of the terrorists was not established, the Pakistan-based groups Lashkar-e-Taiba and Jaish-e-Mohammed were named by the Indian government, and the Indian press published calls to finish this long quarrel with Pakistan once and for all with a full-scale war.

In Pakistan, President Musharraf condemned the attack and banned the two groups, closing down their offices and arresting dozens of their members as well as other extremists. But in the following weeks tensions between the subcontinent's hostile neighbors heightened, and India began to lay mines along the border. The future of Kashmir was once again reduced to a poisonous contest between these rival nations. Each claimed to have the loyalty of the Kashmiri people and blamed the other for the conflict. In this deafening exchange, the voice of Kashmir was silent.

2.

Ramesh Mahanoori is a Pandit, a Kashmiri Hindu who, like most of the half million people in his community, lived until 1990 in relative prosperity in the Kashmir Valley. The Pandits formed the backbone of the professional class, and filled most of the teaching and government jobs. They had deep roots and high status in the valley, and lived side by side with Muslims, sharing the Kashmiri traditions of song and poetry, eating in each other's houses, sometimes worshipping at common holy sites. The harmony between the communities, so distinct from the tensions and violence elsewhere in India, was part of a general culture—the so-called Kashmiriyat—of which both Hindus and Muslims were proud. Even when Partition unleashed mass murder between Muslims and Hindus elsewhere, in Kashmir neighbors of different faiths preserved their courtesies and communal tolerance.

But between January and March of 1990 that tolerance ended, and a quarter of a million Pandits fled the Kashmir Valley, driven out by murders, riots, and death threats. The Pandits had become early victims of the new Muslim insurgency.

Ramesh Mahanoori was once a teacher in the Kashmir Valley. Now in his fifties, he received me in a tiny one-room house in a refugee camp on the outskirts of the city of Jammu. A large bed took up much of the room. There was a sliver of living space where a child sat on the floor, bent over a book; behind a curtain was a crude kitchen where Mr. Mahanoori's wife could be heard preparing tea. A tap outside served as the bathroom. We sat, cross-legged, on the bed, along with two of Mr. Mahanoori's friends. As we talked, his wife appeared and burrowed beneath the bed.

"That's where we keep the stores," he explained.

For Mr. Mahanoori, the expulsion of the Pandits was a straightforward case of betrayal. It began, he believed with the Islamist underground, financed by the I.S.I. Its leaders had started organizing in Kashmir in 1986, and after the farce of the 1987 elections their followers increased. In 1989, an orchestrated campaign of executions began. "The first assassination was of a lawyer," Mr. Mahanoori said. "It was followed by other killings—three hundred and ninety highly selective killings of doctors, engineers, educators, judges. All labelled Indian agents. All our intellectuals." The government, he said, gave no protection. "There was a clear message from the majority that they could no longer live with the Pandits. The Muslims were all united under the banner of *azad*—freedom. Pakistan was their mentor."

Warnings were posted that Pandits who remained in the valley would be killed. Muslim activists set businesses on fire as the police stood by. Fear gave way to panic, and families began to leave. There were rumors of death lists in the local mosques. Pandits throughout the valley hastily packed and fled. Many ended up in the Hindu-dominated security of Jammu, in the south, imagining that they would return in a few months. But after they left their property was looted. Twelve years later, most of them are still refugees.

I heard endless variations of the Pandit story. Some people believed, genuinely, that the assassinated men had been agents of the Indian state. Others believed that the violence had been orchestrated from New Delhi (thus the lack of official interference); this way, the Kashmiri insurgents could be condemned for ethnic cleansing and dealt with accordingly. All versions agreed that the expulsion was brutal, sudden, and comprehensive.

Mr. Mahanoori recalled that in the village where he grew up in Kashmir he had been surrounded by members of his extended family. In the flight, they have scattered, and they rarely meet. "We had the same surnames as the Muslims," he said. "We were all related. They just converted to Islam—only three hundred years ago. Our cultures resembled each other. Here, in Jammu, we are aliens. We have nothing in common with these people."

Now there is a generation of children growing up in a world bounded by the camps. For them, Kashmir is a name, the source of their parents' sadness, something that marks them as different from the people of Jammu. They no longer speak the Kashmiri language, Mr. Mahanoori said. He longed for war. India, he said, should go to war with Pakistan, to resolve this issue once and for all.

3.

As I flew to Srinagar, I had few fellow-passengers—some Indian military personnel, a handful of Kashmiris, and one other foreigner. Even the most intrepid trekkers now prefer to explore other, less dangerous mountains, and the tourist trade that used to sustain the economy has dwindled.

The road into the city was an obstacle course made from an eclectic selection of barriers: metal bars set with eight-inch-long spikes, rolls of razor wire, and oil drums filled with concrete, which forced cars to weave a slow slalom path between them. Each barrier was guarded by men with automatic weapons. Beside one, in a bizarre juxtaposition, a poster offered a seductive welcome. "Kashmir—an adventure," it said. "The land of forests."

The light was fading as I reached the last barrier before my hotel, which had been recommended as a secure place to stay. The driver stopped and switched on the interior light. Beyond my own reflection in the glass, an armed guard was peering suspiciously into the car. In the deserted lobby, Muzak was playing to empty armchairs. Three men looked up from the reception desk in surprise. The lobby was so cold that I could see their breath in the dank atmosphere.

India has now fought three direct wars in Pakistan, two of them over Kashmir. For India, the insurgency that began in the late eighties is another war with Pakistan—a proxy war, in which the enemy is Pakistani-trained infiltrators, with weapons and money supplied by the Pakistani intelligence services. This invasion of its territory, India argues, is a straightforward attack on its sovereignty, and demands defending. For India, Kashmir's status is incontrovertible: it is a part of the Indian state, a senior official told me, and there is no negotiation on either sovereignty or territory. India's response, therefore, has been a military one. But the nature of that response has created a conflict with the wider population of the Kashmir Valley.

Early in the morning on January 13th, four days before I arrived in Srinagar, two men were shot dead by Indian security forces, on a road near Dal Lake. Their names were Ahmed el Bakiouli and Khalid ed Hassnoui, and it was reported that they were foreigners who had attacked a Border Security Force patrol. In the ensuing incident, the two men were fired on by soldiers on watch in a fortified bunker nearby. By the time local photographers arrived, Ahmed and Khalid were dead.

The fact that Indian soldiers had shot two men was not in itself newsworthy. Since 1947, India has maintained a heavy security presence in Kashmir, one that is now half a million strong. To the local Kashmiris, these forces look and behave like an occupying army. With the exception of the local police—who are regarded with suspicion by the Army and the paramilitary forces drawn from elsewhere in India—few of these forces speak Kashmiri. They, in turn, are far from home, surrounded by people whose language they cannot understand, and threatened by an enemy they cannot identify. To the Indian security forces, anyone they encounter could be a terrorist infiltrated from Pakistan. "If a dog barks in the market," one trader told me, "the Indians call him a Pakistani."

The men the solders are looking for belong to any number of dangerous militant groups, many with competing objectives—some wanting independence from India, or an Islamic state, or a union with Pakistan. Several groups began to impose a more severe version of Islam on the tolerant culture of the valley: women were made to wear veils, and bars and beauty parlors were closed down. Foreigners were attacked. In 1995, a Pakistan-based Islamic rebel group kidnapped six Western trekkers: four vanished and one escaped; the sixth was decapitated. There were plane hijackings, which sometimes led to the release of captured terrorist leaders; armed encounters in the mountains and villages; and car bombs and grenade attacks in the cities.

Indian security forces responded with repressive tactics. Shopkeepers and university professors, impoverished farmers and well-heeled businessmen continue to complain of routine cruelty exercised by the security forces during cordon searches: entire districts are sealed off, and the inhabitants are turned out of their houses and made to squat in the cold for hours as the troops ransack their homes. Men and boys are beaten; there are shootings; valuables go missing.

> "We are keeping them safe," an Indian officer said of the Kashmiris. "You never know which vehicle a terrorist might be driving."

For the Indian security forces, such operations are a necessary part of a war against an unseen enemy—

one who might be disguised as a market trader or as a schoolboy or even, as in the case of Ahmed and Khalid, as a pair of out-of-season travellers. But this time it was not just the people of Srinagar who were skeptical of the official account. Ahmed and Khalid, it emerged, were neither Kashmiri nor Pakistani. They were Dutch nationals of Moroccan descent who had ostensibly come to Srinagar as downmarket tourists. They had valid travel documents, had signed in at the Foreign Registration Office in Srinagar, and had been spotted at the Tourist Reception Center by a rickshaw driver named Amin Bakto, who was there looking for business. Bakto had invited them to stay at his houseboat, and they had been there for a week, when, according to an inspector general of the Border Security Force, they had gravely injured two of his men in an unprovoked terrorist attack.

Amin Bakto's houseboat, the Happy New Year, sits in a dirty side canal, greasy green water lapping against the boat's peeling paint. Bakto is a small, spare man, and he talked in nervous bursts, as though he were unable to shake the apprehension that he might somehow be implicated in the events that had led to the deaths of his paying guests. He lives with his family on an adjacent houseboat, and was willing to show me where the two Dutch nationals had slept for the week that they had been his guests.

It was a small room that contained little more than a double bed, which they had shared. Gaping holes in the floor were the consequence of the police search that had followed the killing. The room smelled of stagnant canal. Ahmed and Khalid had paid him two hundred rupees a night (about four dollars), Mr. Bakto told me, a sum that included the use of a heater and breakfast, which he had served himself at nine o'clock each morning. The men were pleasant and quiet, he said, and occasionally played with his children. He never saw them pray or visit the local mosque.

On the day they died, he had gone to offer them breakfast as usual, but found that they had left. The Happy New Year was empty, the doors and windows open. The men had set out along the towpath to a nearby road. By 7:20 A.M., they were both dead, sprawled some twenty yards apart on a road now spattered with their blood. Later, the police had found on the houseboat the packaging to a pair of large kitchen knives, apparently bought in a local bazaar. The knives, in the police version, were the evidence that connected the two Dutchmen to a network of international terror.

No local witnesses came forward to corroborate the security forces' story, and almost nobody I met believed the account. The version favored by the local newspapers was that the patrol had been abusing a local woman, and the two Dutchmen had attempted to intervene. Others believed that they had been challenged by the patrol on their way back from morning prayers. They might not have understood an Indian soldier's command to halt. In either case, they risked being shot.

I went to visit Inspector General Gill, who commands the Border Security Force in Srinagar, and whose men had killed the Dutchmen. I had met him on my first evening in town at a rather stiff party attended by the local commanders of the security and intelligence forces in the district. He had seemed cultured and courteous, and it was difficult to connect him with the acts of torture and repression blamed on the men he commanded.

He had suggested that we meet at his bungalow in a hilltop compound that houses government servants. At the first barrier, my car was searched, and the driver and I were body-searched. At the second barrier, we were assigned guards to take us through the third barrier, where an armored car was parked, guarding the approach road. The final barrier was beside the compound gate. From there, I walked to the house. We talked in a small, bare sitting room, warmed by a large metal stove that crackled in the corner.

Gill is a slim man of fifty-one, a Sikh from the Punjab. For him, there was no doubt that the two Dutchmen were terrorists. Their attack, he said, had been unprovoked. They had inflicted eight stab wounds on his men before they were shot; one of his men lost an eye. For Gill, the Dutchmen reinforced his conviction that the war in Kashmir was sustained from outside—a Pakistani proxy war. He admitted that there was no evidence of a Pakistani connection in the Dutchmen's case. "Do I have to prove that everyone has a past career?" he asked me plaintively. He held to his general point: If Pakistan, with its connections to international terror, would stop sending militants into Kashmir, the trouble would subside overnight. It was a conviction that was shared by the Indian government and widely reflected in India's national press. Besides, he insisted, his men did not shoot people without cause, and the many allegations of torture and disappearance made against his forces were scrupulously investigated. And almost none, he said, stood up.

The local press was unconvinced, even though it had been thoroughly briefed on the incident by Gill himself. The widely held feeling in the valley is that the insurgency is no longer masterminded from Pakistan or anywhere else: the native-born movement is now well established, after years of Indian abuses. And that feeling was reinforced by the killing of the Dutch tourists, regardless of what actually happened. Perhaps the men, armed only with kitchen knives, attacked a military patrol. But the belief is that this army of occupation can shoot anyone it wants to, anytime, with impunity.

I tried to explore the region around Srinagar. The roads to the border, where Indian and Pakistani troops continued to exchange mortar fire, were blocked with snow. To travel outside the city was dangerous. The splendid Moghul fort perched on a hill above Srinagar was occupied by the Army. I found myself circling the city, trying not to feel caged. The streets were wet and muddy, with piles of dirty snow. The light was flat and weak, filtered through a morning fog that rarely dispersed during the day. As I drove around, the sense of military occupation was oppressive. On every street, people were being stopped and searched by Indian soldiers, taxi-drivers opening the trunks of their cars for inspection, lines of bus passengers waiting to be frisked. My car was frequently stopped, my documents inspected, and my driver closely questioned. I was harangued by Indian soldiers who considered the stamp on my press pass insufficiently clear.

In the evenings, Srinagar was subdued, and the tension was unmistakable. As night fell, the people I met and talked to often began to fidget, caught between the obligations of hospitality and their anxiety that I leave before the streets became unsafe. Nobody, they told me, goes out after dark. A tremulous sociology professor described to me the social effects of the long war—migration, unemployment, broken families, a startlingly high rate of suicide.

"It's the constant fear," he said. "Torture, tension. Even at home, the security forces can arrive at any minute. We used to be a leisured people. Now all our entertainment has gone. It's out of the question to go out."

Education had deteriorated in the wake of the Pandits' departure, he told me; young women cannot find husbands, married women are widowed and destitute. He urged me to walk around the old city, a district I had been warned against, to discover how people really felt. It was a hotbed of militancy, I was told, and subject to constant cordon searches.

"Talk to people," the professor said. "No one will harm you." After a pause, he seemed to think better of his assurances. "Don't tell anyone in advance. Don't make an appointment, in case, in their innocence, they tell someone you are coming. And don't stay more than half an hour in the same place."

It was now dark, and he was agitated. I drove through the rapidly emptying streets, ready for another evening of chilly confinement in my hotel. But that evening I had a visitor.

I had called Commander Chauhan, of the Border Security Force, several times, and now he appeared, exuding friendly confidence, eager to show me the sights—at 9 P.M., long past the hour when civilians had abandoned the city. The Commander was a portly, bespectacled man dressed in a camouflage jacket and a black beret, and carrying a polished swagger stick and a walkie-talkie. He bustled jauntily into the hotel's freezing dining room and greeted the waiters by name. The waiters smiled anxiously.

The Commander was eager to stress that he had excellent relations with the local people. His job, he said, was to protect them from the militants. His unit had adopted a girl who had been attacked with acid by Islamic fundamentalists for failing to wear a veil. There were orphans whom his men took care of. And, he assured me, they were steadily weaning the Kashmiris off Islamic extremism. Normality, he announced, was visibly returning.

"You see girls driving cars, boys and girls on motor scooters, going out to the lake, going to hotels, cinemas, and beauty parlors," he said. I had seen none of those things.

The security forces, as he described them, were dedicated to social welfare: "If someone's wife goes into labor in the evening, they just ring up. I send a car to take her to hospital. I have had so many calls to say thank you."

I didn't doubt it. Anyone moving around the city at night without military protection, pregnant or not, risked being detained as a terrorist. As though reading my thoughts, Commander Chauhan suddenly said, "What have you seen? I wish we had met earlier. I could have taken you on the lake. We have motor launches, you know. I'll take you out now, to see the city."

Outside, his jeep backed slowly to the door, as six soldiers armed with automatic weapons walked alongside. I climbed into the front, the Commander took the wheel, and the soldiers jumped into the back. Another vehicle moved up behind us. We then pulled out into the deserted street and embarked on a tour of the city.

"This is the polo ground," he said. He waved a hand vaguely at the darkness. "But they don't play much polo these days. And here—this is the golf course. Excellent golf." We drove on in the empty street. "This is the canal." We turned toward the lake. "Have you seen the Nishat Gardens?" he inquired.

Before I could reply, I was blinded by the beam of a searchlight which had appeared from inside a bunker. The Commander braked sharply. A man jumped out from the back of our vehicle and explained our presence to a group of nervous soldiers whose guns were trained on him and on us. Satisfied, they allowed us to pass, and the Commander continued with his description of the delights of boating on Dal Lake. I found myself recalling an incident described to me by a lecturer in the English department of the Srinagar university. A group of graduate students doing research on the lake one day were shot dead in their boat.

A few hundred yards further on, we found ourselves inching around oil drums and rows of spikes in the road, and came to a stop before a final improvised barrier of rocks—and another checkpoint. At the next bunker, however, we failed to stop in time, and there was a fusillade of hostile shouts. Commander Chauhan braked violently. A soldier in the back climbed out, his hands raised, and stood for several minutes in front, in the glare of the headlights, trying to talk down the guard whose gun was trained on him. I held my breath. Five more guns were pointing at the jeep. Even Commander Chauhan had fallen silent. At last, the soldier gradually lowered his arms.

Commander Chauhan had lost a little of his bounce. For the first time, he seemed to feel that he should acknowledge the surreal character of a city tour in which even a senior officer risked being shot by his men. "Actually," he said, "we give them orders that every vehicle must be stopped. You never know which vehicle a terrorist might be driving." As he picked up speed, his faith in his mission returned. "We are keeping them safe," he said. We turned back through the deserted streets of the old city. There was hardly a light showing, but for Commander Chauhan this didn't mean that people were afraid to go out. "Look," he said triumphantly, "they are all in bed with their wives and their blankets, and my men are out here, keeping them safe!" Suddenly, he turned to me. "What did people tell you, by the way?"

"They said they wanted independence," I replied bluntly. "That they were afraid of your men and their searches." I stopped short of telling him how many rejoiced when Indian soldiers were killed.

"Did you ask them why we search them?" he said.

I had not, of course, though I could imagine Chauhan doing so, chattering as he frisked people, in an exercise in hearts-and-minds didacticism, cheerfully explaining his motives as their humiliation deepened.

I asked him how many terrorists he thought there were.

"Very few, these days," he replied.

Why, then, did the government need to keep half a million men here?

"Because," he replied quietly, "you don't know who they are."

4.

The conflict over Kashmir has entrenched the worst suspicions that India and Pakistan nurse about each other. For India, the separate Muslim state of Pakistan represents a rejection of the secularism that India believes to be essential to keeping its own rival religious communities at peace. If Kashmir's Muslims were to join Pakistan, what signal would that send to the more than a hundred million Muslims elsewhere in India? For Pakistan, India's refusal to allow Kashmiri Muslims to join the Pakistani state merely confirms its conviction that India never abandoned a long-term ambition to establish Hindu domination on the subcontinent, or that it even accepted Pakistan's existence. But for the people of the Kashmir Valley, with their distant dreams of independence, neither neighbor offers a solution. Pakistan's muscular Islam is at odds with Kashmir's Sufi-inspired traditions. The Muslims and Pandits of the valley speak a different language from the language of India or Pakistan. Neither country is home, and each, in turn, has been a threat: after all, it was the incursion of tribal raiders from Pakistan in 1947 that brought Indian troops in retaliation.

Thirteen years into the insurgency, the local politicians in Kashmir, like the competing militant groups, have conflicting objectives. Even members of the All Parties Hurriyat Conference, which was formed in 1993 by more than thirty political parties to act as a voice of a people who felt themselves disenfranchised, are quarrelsome and deeply divided. I met many members, and asked them what they wanted for the country. I got many different answers. One wanted union with Pakistan. One wanted independence. Others would be happy with real autonomy within the Indian state. Some had links to the militants; others did not. Each claimed to represent a general majority.

From civilians I got a different picture. Weary of war, few believed that anything could be won, now, through the armed struggle. But few supported Gill's contention that the hostilities would end once Pakistan stopped supporting them. I understood this view when I met some of the young new recruits. There seem, potentially, to be an endless number: the war's capacity to create new militants is limitless.

Militants are buried in Srinagar's many martyrs' cemeteries, some of them large adjuncts to regular graveyards, others crammed into small corners across the city. They are crowded with almost identical concrete gravestones, covered in Arabic inscriptions in green lettering. A few bear English place names—"Birmingham" was one I noted—an indication of Kashmir's appeal to disaffected Western-born Muslims looking for a cause.

But exactly what that cause was, beyond the single word *azad,* was unclear. As I looked at the gravestones in one of the smaller cemeteries near Dal Gate one day, a group of boys gathered around me and laboriously translated the inscriptions. They were eager to show me significant graves—Islamic warriors from faraway countries, or men whose spectacular deaths had stuck in their memories. They pointed out professionals—lawyers, teachers doctors—and men who had died under torture. They called all the dead "martyrs," but they couldn't always tell me what they had died for—whether they were martyrs to Kashmiri independence, or to the union with Pakistan, or simply to Islam. One grave that was pointed out to me belonged to Aafaq Ahmed Shah, who, at the age of eighteen, had become briefly famous for inaugurating a new phase of the militant group Jaish-e-Mohammed's war, that of the suicide bomber. I had earlier visited his family in the old city.

His mother had opened the door and stood on the doorstep looking at me. She was a small, middle-aged woman with dark circles under her eyes, and she knew, before I explained, what I had come for. She remained immobile, tears flowing down her face, reluctant, it seemed either to turn me away or to admit me to what she knew would be a painful rehearsal of her grief.

Still sobbing, she let me in, and I sat on the floor of a freezing room, waiting for her husband to return from the market. Mohammed Yusuf Shah was a thin, elderly man, a retired college teacher dressed in a brown *pheran.* He settled beside me as his wife brought blankets and fire baskets and poured us cups of chai from a thermos flask.

"My son was nearly nineteen years old," Mohammed said. "He wanted to be a doctor. There's a photograph of him"—he waved his hand vaguely—"somewhere, wearing a stethoscope." He made no move to get it, as though already discouraged by the effort. His wife had begun to cry again.

"Mysterious are the ways of God," he said. There had been no warning that his son would join the militants. "He willed it. He did it. That is all. He was a good, silent, obedient boy. He was my son, but, more than that, he was my friend. He was here, dawn to dusk, every day, day and night."

On March 25, 2000, the boy disappeared. The family searched for him, fruitlessly. Three days later, he telephoned. "Father, I left," he said, and hung up.

On April 19th, dressed in an Army uniform and carrying an Army I.D., Aafaq tried to ram his car through a heavily fortified gate of the Army's XV Corps headquarters, near his home. The car exploded after a solider started firing at it. Five soldiers were injured; only one person was killed—Aafaq, who was blown to pieces. The family read of his death in the newspapers.

I found another family of a young martyr in a village some thirty miles outside Srinagar. We drove along long straight roads lined with tall poplars, past fields of saffron that were just showing a first flush of green shoots, past empty paddy fields, waiting to be planted. The village itself was along a muddy track, buried among trees, peaceful in the chilly morning. "Ignorance is the root of all evil" was carefully painted on the wall of the village school.

I sat on the floor of the family's small living room as villagers crowded in, competing to tell the story of Nazir Ahmed Khan.

"Nazir was in the tenth class," a young neighbor told me, and also wanted to be a doctor. "His hobbies were gardening, photography, and cricket."

Nazir's elder brother, Mohadin, drove a taxi to support the family. There was a skirmish in a nearby village, and the Army appeared at the door, convinced that Mohadin had driven a wounded terrorist to the hospital. Mohadin was not at home, but Nazir and his father were. They were interrogated, but the soldiers were not satisfied, and the father and son were both beaten, and then their limbs were held over a fire. Afterward, Nazir ran away. "He could not bear being tortured for no reason," the neighbor said. He had gone to join the militants.

Mohadin was summoned to the Army barracks, and he, too, was tortured and then imprisoned. The family sold the taxi to bribe the Army for his release; it was their only asset. And then Nazir was killed.

I went to see where he had died. We drove back to the main road, past a sign that read, "Our job is to make everybody see the beauty of Kashmir," then turned down a muddy track to the village. We inched along the narrow streets until a villager pointed to the house: the roof at one end had collapsed, and its supporting wall was a pile of rubble. I scrambled up the slippery lane and pushed open the ramshackle corrugated-iron gate. A small crowd followed me in.

The boy had joined two other militants, and the three of them, the villagers told me, were hiding in this house. An informer told the Army. The cordon search lasted for three days and three nights, and the entire village was made to squat in the cold on the recreation ground. Fire baskets and the old men's woollen hats were confiscated. Ten thousand soldiers came, I was told. I said that ten thousand soldiers is a very large number. The villagers insisted.

Cornered, the militants gave themselves away—one of them fired on the soldiers from an upper room—and the Army ordered seven villagers to walk up to the house and put two explosive devices inside. Everyone knew that the villagers would be harmed. It was, they said, a frequently used tactic. "The militants don't fire on civilians," a villager explained. "If you refuse to do it, the Army shoots you." The villagers got out before the devices were detonated. Nazir was eighteen, and had been a militant for a week.

Later this year, there will be elections in Kashmir—the opportunity, in principle, for the people to express their political will. But, after years of vote rigging and intermittent direct rule, Kashmiris have lost their faith in India's secular democracy. For the politicians in the loose coalition of the All Parties Hurriyat Conference, there is no point in even standing. To do so would require their swearing an oath of allegiance to the Indian state, which they do not wish to honor. And, at the very least, they want an autonomy that they believe India's current government—dominated by the Bharatiya Janata Party, an organization with an aggressively pro-Hindu ideology—will never grant. On February 12th, the Hurriyat announced that it would boycott the Indian elections and hold an election of its own—to choose representatives who will sit at a negotiating table with India and Pakistan. But India is not going to give up Kashmir, and the negotiations have no hope of succeeding.

President Musharraf, too, has called for negotiations, and on February 6th the U.N. Secretary-General, Kofi Annan, offered to mediate. For Musharraf, negotiations could be the key to his survival. He has declared his wish to make Pakistan a more secular state, attempting to dismantle the networks of Islamic extremists who, for the past twenty years, have systematically infiltrated Pakistan's government, Army, and, especially, its intelligence services, the I.S.I. These people are viscerally opposed to Musharraf's ambition. If he is to succeed—if, at the very least, he is to put an end to the I.S.I.'s support for cross-border Islamic terror—he needs to show that the cause of the indigenous Kashmiri struggle has not been abandoned. India, meanwhile, has not taken up the offers for negotiations.

In Kashmir, an end to the struggle seems ever more remote. Nazir, like Aafaq, had joined the ranks of the martyrs in a war that has lost its way, a war that now feeds on itself—each act of violence generating a new response that generates more recruits. For some of the valley's young men, it can seem as though there were little else to do—the war is their occupation. The Kashmiriyat is now a forlorn memory, and has been replaced by the cult of the gun. The people of the valley believe they are trapped in a war without end, in which anyone can become a victim. Tens of thousands have died. Scarcely a family in the valley is untouched.

The branding of Hong Kong

A dragon with core values

Stick that in your lapel

HONG KONG

ON HANDOVER day in 1997, Donald Tsang, then Hong Kong's finance secretary, pinned a little emblem on to his lapel. It was a double flag—Communist China's joined to Hong Kong's *bauhinia* flower—that stood for "one country, two systems". For more than three years Mr Tsang was rarely seen without it, and it became, along with his bow ties, his trademark.

Last year—by when he was chief secretary—Mr Tsang replaced the emblem with a little dragon, the fruit of three years of research by international brand consultants. Besides cosmopolitanism, says Kerry McGlynn, the government public-relations director behind the project, the dragon projects five "core values". These are three adjectives—"progressive", "free" and "stable"—and two nouns, "opportunity" and "quality".

The visual link, according to the government, is self-evident: Hong Kong stands for "East meets West". So the dragon is composed of two parts that could, if you twist it, stand for the letters H and K, as well as the Chinese characters *Heung* and *Gong*. Combined into a dragon, an ancient Chinese metaphor for energy, the strokes represent Hong Kong's legendary dynamism.

The dragon appeared on brochures, buses and much else last summer, and within days Hong Kong's people were naming it. Expatriates saw it mostly as a "flying fox", while Cantonese speakers—usually more creative in such matters—settled on "shocked chicken". Those appraised of the consultants' fees called it "the HK$9m dragon".

Perhaps the most perplexing thing about the dragon, however, is that it took Hong Kong so long to get one. Canada branded itself in 1970, and New York ("the Big Apple") a decade later. Besides, their fetish for brands is one of the few core values that most Hong Kong residents agree on. As one long-term resident puts it, "When the going gets tough, Hong Kong goes shopping."

L.A. Area Wonders Where to Grow

Development plans raise fears of more traffic and pollution as metro rethinks growth direction

By John Ritter, USA TODAY

LOS ANGELES—Plans for two of the biggest housing developments ever built here in subdivision paradise make James Chang wince.

Chang and his wife, Kelly, epitomize the postwar Los Angeles story: young, upwardly mobile suburbanites chasing dream houses in the ever-expanding, auto-reliant megalopolis. The Changs camped outside a homebuilder's trailer for five days and nights to land—sight unseen—a new, $500,000 four-bedroom house. Never mind Chang's hour-and-15-minute commute to west Los Angeles. The same house close to his job would go for at least $1 million.

The Changs wanted good schools and a safe neighborhood and found them in Santa Clarita, a city of 163,000 up "the five"—Interstate 5—from the San Fernando Valley. But now Chang dreads an even longer commute if the two big developments add 140,000 people to the thinly populated northern edge of sprawling Los Angeles County.

"The traffic is just going to be horrendous," worries the law enforcement officer, 37.

Arid high desert north of the San Gabriel Mountains is the county's last big chunk of developable open land after decades of relentless growth beyond the coastal core. In rush hour, the commute downtown is a minimum 90 minutes. If developers' plans move ahead, opponents say congestion and pollution will worsen, environmentally sensitive landscapes will be paved over and water supplies will become more strained than ever.

The two proposals—one approved, one facing environmental studies—have stoked a debate over the future of the nation's second-most-populous metropolitan area. A culture revolving around the auto and the single-family detached house, with mass transit a Johnny-come-lately, is forced to rethink whether that ethic can prevail much longer.

Running Out of Land

Los Angeles isn't the USA's only growth-constrained region. Miami is out of land and hemmed in by the Everglades. Las Vegas, the nation's fastest-growing metro area, soon will hit a barrier of federal land holdings—if it doesn't run short of water first. Less confined but still facing growth challenges are Seattle, Phoenix, Portland, Ore., Houston, New York and Boston.

Either by choice or because they must find affordable housing, people in many locales are moving farther and farther out, accepting longer commutes for the chance to buy a house.

"The message is, given our country's bounty of land, some of its bigger metro areas have bumped into limits that would have seemed impossible to our parents and grandparents," says Robert Lang, director of the Metropolitan Institute at Virginia Tech.

When Los Angeles' boom began in the 1920s, few thought its huge coastal shelf would ever be built out. But suburbs spread to the foothills and raced through the canyons and into the mountains as far as they could go.

Lang finds the change ironic. "It's at the core of the California being to have freedom of movement, freedom of space," he says."But those are the very things at risk and certain to change in the next 20 years."

The five-county Los Angeles metro area will add more than 5.3 million people by 2030, the Census Bureau estimates. Los Angeles County alone will add 2.4 million. Where will they all live? Will subdivisions keep spreading into the desert? Could Southern California become one long urban traffic jam stretching more than 200 miles from San Diego to Bakersfield?

The answer is that it's a region in transition, growing differently as suburbanization comes full circle. The trend now is to revitalize the city and older suburbs. Mass transit could never compete with the almighty car, but now light rail lines are magnets for commercial and residential projects. So are scores of

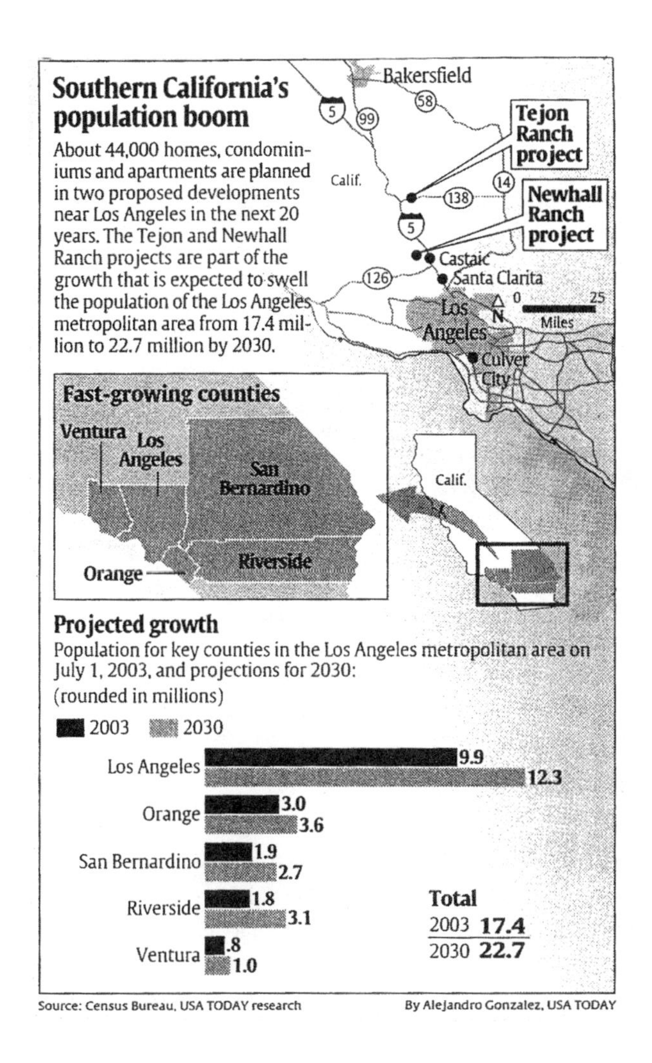

aging strip malls. The trade calls that "infill"—filling in vacant properties and redeveloping others.

"L.A. is about to become the world's biggest experiment in infill development," says William Fulton, president of Solimar Research Group and a city councilman in nearby Ventura. "There's decades of potential in the obsolete areas from the '60s and the '70s."

Developers who saw no market a decade ago are rushing to cash in. "Most of them realize this is where the opportunity is because the old game is just about played out," Fulton says.

Town houses and high-rise condos and apartments will be in demand. But consumers' first choice—single-family detached houses—will be harder to find in the city's core.

Projects that cut auto trips are in vogue. "We don't have the luxury anymore of having a retail center in one location and housing in another," says Rick Caruso, chief executive of Caruso Affiliated, a development company that has six infill projects underway and more soon to be announced.

Infill will be tougher and costlier because of NIMBYism—Not In My Back Yard—but Los Angeles has no other options.

The two big north-county projects, Newhall Ranch and Centennial, are a dying breed in the era of great Southern California planned communities.

Environmentalists fought Newhall Ranch for years until a judge last fall cleared the last legal hurdle to a plan for nearly 21,000 homes in scenic Potrero Canyon.

"The bottom line is we have planning by developer while the river develops arteriosclerosis. It's atrocious, and the taxpayers pay for it over and over again."

—Lynne Plambeck, president of the Santa Clarita Orgainzation for Planning and Environment

Opponents intent on saving farmland, keeping the Santa Clara River healthy and protecting endangered species such as the unarmored three-spined stickleback, a tiny silver fish, say they'll contest subdivision schemes as they come up for review.

"The bottom line is we have planning by developer while the river develops arteriosclerosis," says Lynne Plambeck, president of the Santa Clarita Organization for Planning and Environment. "It's atrocious and the taxpayers pay for it over and over again."

A Battle Over Water

Centennial, 23,000 proposed homes on 12,000 acres in the vast Tejon Ranch 60 miles north of Los Angeles, appears more sticky for developers. Opponents say the land is critical for wildlife, including the endangered California condor, because habitats of two deserts and four mountain ranges converge there.

Centennial project manager Greg Nederios says the Tejon Ranch site has been grazed for 150 years, most of its native grasses are long gone and no threatened plant or animal species have been found, though independent studies have yet to confirm that.

Zev Yaroslavsky, the only county supervisor to vote against Newhall Ranch, says more sprawl and longer commutes are taking a toll. "It's not good for the air quality of our region. It's not good for the social stability of our region. At some point, enough is enough," he says.

Marco Marenghi, 37, a computer animator for Sony Pictures in Culver City, recently bought a house in Castaic, up I-5 from Santa Clarita. He likes it that Newhall Ranch will attract more retail services, but "the downside is traffic." By car, Marenghi's commute is 90 minutes. "I tried that once and hated it," he says. "Now I commute with a motorcycle, and I can do it in 45 minutes."

REINVENTING A RIVER

Its waters drove our first Industrial Revolution—and were poisoned by it. Thoreau believed the Merrimack might not run pure again for thousands of years, but today it is a welcoming pathway through a hundred-mile-long red-brick museum of America's rise to power.

by Cait Murphy and Rosanne Haggerty

MATTERS DID NOT LOOK PROMISING. THE PATH DOWN TO THE canoe launch onto the Merrimack River was long and steep, thick with roots and brambles and sharply angled. Pushing, pulling, and grunting, we reached a scum-slicked spit of sand just below a wide stretch of renovated nineteenth-century mill buildings in Manchester, New Hampshire, and pushed off.

In other words, the Merrimack does not always make it easy for boaters to play on its surface. But that is not surprising; until quite recently no one would have wanted to. Its image problems go back more than a century. Described by *National Geographic* in 1951 as "a veritable slave in the service of industry," this 116-mile-long river was dammed, canalled, and dumped on to within an inch of its life. Longtime residents recall watching what they sometimes called the "Merrimuck" change color depending on which dyes the textile mills were using that day. Its vegetation grew in mutant forms. As a repository for everything from medical waste to offal, it reeked.

Why bother with the Merrimack? Because it encapsulates much of New England's history, colonial, industrial, and postindustrial. Because Henry David Thoreau traveled along it to write his elegiac *A Week on the Concord and Merrimack Rivers*. And because it is beautiful.

The waters of the Merrimack powered America's first Industrial Revolution, initially for the textile mills of Lowell, Massachusetts, and then for those in communities like Manchester and Nashua, New Hampshire, and Lawrence and Haverhill, Massachusetts. But just as the river was sacrificed with little regard for its long-term health, so the vast mill complexes died from underinvestment and resistance to change.

The river's comeback began with the Clean Water Act of 1972, which required that sewage be treated before it reached the nation's waterways. "As soon as we stopped dumping pollutants, the river cleansed itself, and much sooner than it was believed it could happen," said Chuck Mower, a local historian and furniture maker. And not just people appreciate the difference. Bald eagles now nest on the Merrimack's banks, as do ospreys, egrets, and hawks. American shad, striped bass, trout, and Atlantic salmon swim in its depths. Otters and minks frolic on its shores. What happened was an unusual symbiosis. Communities acted to regenerate the river, an effort that today embraces hundreds of volunteers who monitor its temperature, pulse, and respiration, and the river returned the favor, helping regenerate the communities on its shores.

The Merrimack officially begins where the Winnipesaukee and Pemigewasset Rivers join in Franklin, New Hampshire, the birthplace of Daniel Webster, whose home you can visit. Determined to start at the exact beginning of the river and assured that we couldn't miss it (which naturally made us nervous), we parked downtown and asked around for the source of the Merrimack. We might as well have been asking for the source of the Nile. Before long, though, we think we have found its origins behind a parking lot.

For a body of water once so famously tainted, the stretch from Franklin to Concord is improbably bucolic. Mile after mile there is little to see but birdlife, trees, fields of corn, and the occasional church steeple poking up above the vegetation. The Pennacook Indians, who were there long before the factory builders came, might well find the scene familiar.

We ended our day of paddling in Boscawen, just north of Concord, where perhaps the most politically incorrect monument in America bears witness to the difficult relations between Native Americans and early European settlers. A short walk away from U.S. Route 4, it is a 35-foot-high granite statue to Hannah Dustin, erected in 1874 and the first statue in America dedicated to a woman. Dustin's story began 60 miles away in Haverhill, Massachusetts, which has a more elaborate monument to her. There, on the morning of March 15, 1697, Indians descended on the town, burned half a dozen buildings, and took several people captive, including Dustin, then recovering from the birth of her twelfth child. The baby was killed in the raid. On March 31, held on an island in the Merrimack in what is now Boscawen, Hannah and two companions stole some hatchets and struck back, killing 10 of their 12 captors. They then coolly took the scalps, made their way home, and collected a bounty. Boscawen's statue to Hannah commemorates the spot where she killed the Indians. Stepping forward with a tomahawk in her right hand and a clutch of scalps in her left, she is clearly a woman to be reckoned with.

A bit back down the railroad tracks from her statue, a group of decayed buildings on the left bear witness to another kind of history. They are the remains of a once-thriving flour mill and tannery complex, now being restored, bit by bit, by a local family that has bought the whole site. It is slow work, but you can easily see the attraction. The mass of brick, granite, and wood structures is striking, a living, if disheveled, history of the gritty industrial life that defined New England for more than a century after the first mill opened along the Merrimack in the 1820s.

The river runs through Concord, New Hampshire, but somehow the state capital is not a river town. Concord sits back from the water, and unlike every other city on the route, it lacks massive squat brick mills, favoring instead graceful Victorian architecture and granite buildings designed to impress people with the seriousness of government. The first of the great Merrimack mill towns is Manchester, about 20 miles to the south. On a sunny Sunday in August, the water is so clear we can read the labels on the refrigerators and tires dumped in the past and not yet cleared away; the air is pure, the scene quiet. Even as we skim past a full mile of mill buildings, it just doesn't seem possible that for a century this was a pulsing center of industry.

Manchester grew up around the Amoskeag Mills, which grew up around the Amoskeag Falls of the Merrimack River. In its day the Amoskeag mill complex was the world's largest, employing 17,000 people in the early 1900s. In their uniformity and scale, Amoskeag's buildings resembled a medieval walled city, an all-inclusive social world, as the historians Tamara Hareven and Randolph Langenbach have noted. Starting with its first owners, the Boston Associates (of whom more later), the Amoskeag was always more than a company. It was a way of life. "At one time, the Amoskeag owned practically everything in town," Frederic Dumaine, Jr., who began work at the mills as an errand boy, told Hareven and Langenbach for their oral history of the complex. "The churches and the YMCA received land from the Amoskeag free. All the parks were given to the town by the Amoskeag. They had sewing classes, cooking schools, gardens." By the early 1920s the currents that were to undermine all of New England's nineteenth-century textile production—failure to modernize, labor problems, competition from the South, and excess capacity—had weakened the Amoskeag. In 1936, after a string of punishing strikes, it closed, devastating the local economy. Merchants made an effort to attract small-scale businesses into the abandoned mills, but many of the buildings didn't see a tenant for 40 years, and most of the businesses that did move in were gone by the mid-1970s.

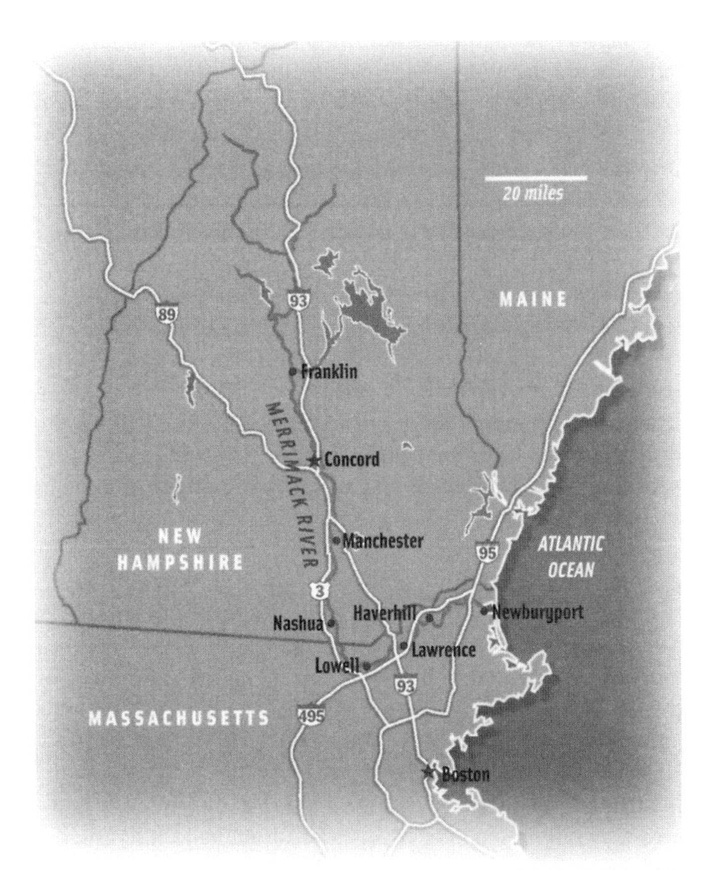

Map by Joyce Pendola

Then things got worse. "The year 1991 was a real low," recalled Manchester's director of planning and development, Robert McKenzie. "On one day, five of the state's six largest banks, which were headquartered here, closed, with the loss of thousands of jobs. It forced the community to come together." A Citizens Planning Revitalization, or CPR, committee was born, involving 200 community leaders in a near-desperate attempt to rescue their city.

Despite its industrial past, fully 80 percent of the New Hampshire portion of the Merrimack watershed is undeveloped farmland, forest, or wetland.

The CPR team had one thing going for it, the power of example. Dean Kamen—a developer of medical devices, promoter of science education, and conceiver of the Segway Human Transporter, a well-hyped campaign to transform life as we know it with a sort of motorized scooter that can think—had turned his visionary eye to Manchester's abandoned mills in the mid-1980s. He bought eight of them and renovated seven. Scorning the use of government incentives ("they add time and money," said his head of property management, Don Clark), he moved his own company, DEKA Research and Development, to the mill yard and attracted such tenants as Texas Instruments and Autodesk. In all, he has reclaimed 750,000 square feet of mill space, and his success has stimulated the city and other developers to rehabilitate the remaining mill-yard properties.

Following more familiar economic development strategies, the CPR created a downtown master plan and a business improvement district and built a civic center. The city has attracted educational institutions like New Hampshire College to the mill yard, established a historic district, and redeveloped a former military base into Manchester Airport. Manchester has clearly come back from its dark days. The downtown is active with restaurants, shops, and cultural institutions, and the mill yards have probably never looked better. Planning for an extended riverwalk is under way, but making the river more of a destination in itself has surprisingly not been a priority. During a full day on the Merrimack near Manchester, we saw no other boaters. Perhaps this is just a matter of time; to many who grew up along it, seeing the river as recreational still doesn't come naturally.

THE BEST WHITE WATER ON THE RIVER SURGES BETWEEN Manchester and Merrimack, which are also rich in remains of the Middlesex Canal, built in 1803, an engineering warm-up for the Erie Canal. The Middlesex was built to bypass the rapids and allow for clear passage of the lumber from northern New England, used by the shipbuilders of Newburyport. Later, during the factory-building era, Merrimack Valley farmers discovered a lucrative sideline making bricks from local clay. But the mills used the bricks to build dams to spin turbines to drive textile looms, and the river's purpose became power, not transportation. Good-bye, canal-boats; hello, railroads.

Paddling south, we are alone on the river, passing farmlands and an island Thoreau once camped on; fully 80 percent of the New Hampshire portion of the Merrimack watershed is undeveloped farmland, forest, or wetland.

That changes as the river runs south, and as early as 1839 Thoreau could see the change coming. Camping outside Manchester on his own trip on the Merrimack, he could not be bothered to note the developing city, but he wrote of Lowell, 20 miles south, "Perchance, after a few thousands of years, if the fishes will be patient, and pass their summers elsewhere... nature will have leveled... the Lowell factories, and the Grass-ground River [will] run clear again."

Lowell's big break came in 1978, when Congress designated it the nation's first urban National Historical Park.

The river is running pretty clear now, but the Lowell factories are unleveled. In fact, they are the centerpiece of a stunning transformation of a decayed New England mill town into something like a showcase. Lowell's modern history began in the 1820s. A hard-eyed business venture from the start, it was nevertheless informed with a visionary idea—that the miseries of industrialism in Europe need not be replicated in America. A group of merchants known as the Boston Associates set about using the Pawtucket Falls, where the Merrimack drops 32 feet, to power a factory complex that, although enormous, would create no permanent working class. The mill hands would be local farm girls who would return to their homes after a few years of wage earning and self-improvement. Good working conditions, albeit for 14 hours a day, a church, supervised boarding homes, a program of lectures and cultural enrichment—all would prove that America could have industrialization without the horrific social effects that came with it in England.

In 1826 Lowell had just 2,500 people and a few looms; by 1850 there were 35,000 residents, and 10,000 workers were producing almost two million square feet of cloth a week. Labor relations were not as smooth as the official story claimed. The "mill girls" struck three times between 1830 and 1840, protesting wages and working conditions and demanding a 10-hour workday, which finally came—for women and children—in 1874. After the Civil War, perhaps weary of dealing with these tiresomely independent Yankee females, the mills' managers turned to immigrants, particularly to the Irish and French Canadians. These newcomers were willing to work longer hours for less money, and they neither required nor wanted paternalistic supervision. The transition to immigrant labor made economic sense, but it also spelled the end of the Boston Associates' grand experiment and the beginning of the industrial working class they had been determined to never let form.

Lowell prospered through the First World War, its population peaking in 1920 at just under 113,000. But then, as in Manchester, the mills began to decline. By the mid-1950s the last of the original ones had closed. "The Boott Mills—the great silent light shrouded the redbrick in a maze of haze sorrow," wrote the Lowell native Jack Kerouac in 1959 of what remained.

Much of the credit for the city's turnaround goes to local leaders like the late Sen. Paul Tsongas, a native of the town, U.S. Rep. Brad Morse, and the educator Patrick Moogan, who imagined that the story of Lowell's beginnings as the nation's first planned industrial city could be the key to its renewal. In contrast to the individual entrepreneurship and locally funded efforts that returned Manchester to its feet, Lowell relied on massive federal and state investments. The big break came in 1978, when Congress designated it the nation's first urban National Historical Park. If you've heard of an economic development strategy, chances are it has been tried in Lowell: downtown facade improvements, streetlamps, plantings, artists' housing, historic districts, tax abatements, public art, a Conference center, plus a rail link to Boston. "The city's needs were such that it couldn't rely on one thing," said Brian Connors, of Lowell's Division of Planning and Development.

Lowell's tenacity and inventiveness is its hallmark. There are exceptional interpretive exhibits at the restored Market and Boott Mills, including a working weave room. There is the recreated boardinghouse for mill girls, the restored locks, the riverwalk that links downtown to the new baseball stadium that hosts the minor-league Lowell Spinners. The "Run of the Mill" canal tour, sponsored by the National Park Service, is a cruise through history along a portion of the 5.6 miles of canals that provided the water that powered the town's mills. And all around there are signs that explain the city's past with a lack of romanticism perfectly in keeping with the down-to-earth character of the place. One such marker, for example, quotes the remarks in 1907 of a factory inspector worried about the dangers of "working days in a room where others in incapacitative stages of consumption (tuberculosis) habitually spit on the floor."

IF YOU CAN STAND A LITTLE MORE HISTORY, CHECK OUT THE American Textile History Museum; the walking tour of the Acre, a neighborhood that began as a "Paddy camp," a tent city for the Irish laborers who came on foot from Boston to dig the first canals; the memorial to Jack Kerouac; and the Whistler House, birthplace of the painter James McNeill Whistler, whose father was Lowell's first rail engineer.

Lowell is not rich; beyond the restored downtown, many a neighborhood is clearly struggling. But it has both found its past and stayed true to it. Still a port of first resort for immigrants, it is home to the world's largest community of Cambodians outside Cambodia.

A few miles downriver the Merrimack takes a sharp northeastern turn and passes through small towns and suburbs on its way to another remnant of the Industrial Revolution. This is Lawrence, which is a kind of anti-Lowell, the image of what Lowell would have been without the national park, Paul Tsongas, and all the rest. Clearly, the comparison hurts. Charlie Boddy, a land-use planner for the city, has to concede that his town, founded in 1845 by the ubiquitous Boston Associates, is younger than Lowell. But in every other way, he insists, when it comes to historical importance, "Lawrence wins hands down." Its dam was the biggest in the world. The clock tower that looms over the city is the world's second largest, just six inches smaller than Big Ben. And the 1912 "bread and roses" strike was a turning point in labor history.

Lawrence *is* impressive. Its old mills line the river as far as the eye can see, and they trace the early designs of its planners. Boardinghouses for workers were sited immediately parallel to the mills (one of which has been restored as the Lawrence Heritage State Park), followed by a row of commercial and municipal buildings and then another of houses of the gentry. But many of the mill buildings are either vacant or used as warehouses; a few downtown streets have been spruced up, while others are desolate. Lawrence, in short, is a work in progress. Still, it *is* progress, fueled by a population that, as before, is composed largely of immigrants, this time from Latin America. "Lowell had a lot of political players with a lot of vision who came on the scene early on," said Boddy. "That is where we can be, given the same amount of time. We're on the same track, just in a different and earlier place."

One of the intriguing facts of life in New England is how towns next door to each other can have such different characters. Haverhill lies on the same side of the Merrimack as Lawrence and just a few miles away, but it's a completely different kind of place. Though for some years it was a major shoemaking center, the "Queen Shoe City of the World," Haverhill looks and feels like the old New England village it still is. Founded in 1640, it enjoys the distinction of having been the inspiration for *Archie* comics (*Archie*'s creator, Bob Montana, was a 1939 graduate of Haverhill High, immortalized as "Riverdale High.")

But it is Newburyport, where the Merrimack opens into the Atlantic, that beckons us on this stretch of the river. Newburyport was never as rough-edged as Manchester, Lowell, Lawrence, or even Haverhill. Still, being downriver from all those factories took its toll. "I remember growing up seeing the scum caused by the bacteria in the water," said Suellen Welch, who works in the harbormaster's office. "Sometimes there would be dead fish on top, and there was a stench." Newburyport also suffered the economic decline all too characteristic of the region.

Yet its history is entirely different. European settlers founded it in 1635, quickly built a church, and then proceeded to engage in three decades of arcane doctrinal disputes. Usually this was civil, always it was earnest, but occasionally it became sinister. Newburyport had tried one of its residents as a witch in 1680. Remarkably, three houses from that era survive: the Noyes House, built in 1646, and the Tristram Coffin House and the Spence-Pierce-Little Farm, both also from the mid-seventeenth century; all are open to visitors.

Newburyport saved the buildings that are its heart, but the simultaneous restoration of the Merrimack gave it back its soul.

Diverting as the religious disputes were, by the end of Newburyport's first century much of its attention had been given over to shipbuilding. The region's pine and oak provided raw materials, and the town's access to the Atlantic made it an ideal location. For a hundred years Newburyport was one of the nation's leading shipbuilders, turning out thousands of vessels. The late 1700s and early 1800s were Newburyport's glory days, as it grew rich on the ships it built and the cargo they brought home, a good deal of it illicitly. Then protectionist legislation in 1807 and 1812 killed off international trade, while the Middlesex Canal diverted traffic to Boston; in 1811 a fire swept through downtown. But the town rebuilt itself, this time in brick, and endured as an important point of entry for foreign vessels. The mortal blow came in the 1840s, when railroads displaced the port. Like many of its upstart neighbors, Newburyport turned to cotton mills and immigrants for its economic survival. When the mills went overseas, or south, Newburyport hit the skids.

BUT LIKE ALL THE TOWNS WE VISITED, NEWBURYPORT NEVER lost its spirit. In the early 1960s, when it had reached its nadir of dilapidation, with much of the downtown boarded up, the town officials took action. They got a plan approved by the federal government: Level 20 acres of downtown, including Market Square, the site of a rebellious tea party in 1773 that may have given Boston the idea, and replace it all with a big shopping center. It was urban renewal, and one stretch of Federal-era buildings gave way to a parking lot.

Then a band of saviors from the morning time of the preservation movement stepped in. They formed a committee and in

1963 invited Mayor Albert Zabriskie to come have dessert with them. "That was a legendary moment for the town," said Mark Sammons, the executive director of the Newburyport Maritime Society. In fact, the committee more or less had the poor mayor for dinner. "We told him we were not happy with what was going on," recalls Ruth Burke, a Newburyport native who hosted the gathering. Then they presented an alternative. Instead of demolishing downtown, why not redeem it? The mayor was won over by a slide show that hinted at just how beautiful Newburyport was beneath the grime, and the feds came on board as well, allowing the town to use the money granted for demolition for preservation instead.

That was only the beginning of what has become a more than four-decade-long labor of love. The results are spectacular. The Custom House Museum, from 1835, offers a sense of the renaissance. At one level it is underwhelming, with not particularly great paintings and artifacts like the jawbone of a whale. But the building itself, designed by Robert Mills, the architect of the Treasury Building in Washington, is a marvel.

The best way to appreciate the place is by walking around it. Newburyport is, quite simply, one of the loveliest towns in New England. What *National Geographic* noted in 1951 still exists—"perhaps the largest and most notable collection of square, well-proportioned, three-story, 19th-century houses to be found anywhere in the country." Street after street is lined with these meticulously restored treasures. The waterfront bustles with boats, and its harbor seals have returned. If Newburyport has any flaw, it's that it might be too perfect. Full of Ye Olde This and Thatte Shoppes, it is the kind of place where a birdbath is easier to find than a hammer. Still, a surfeit of preciousness is a small price to pay for this miracle of preservation. Newburyport by itself could have saved the buildings that are its heart, but the simultaneous restoration of the Merrimack was what gave it back its soul. "The river is the jewel of this city," Welch concluded. "It defines the whole waterfront, which is the grand jewel."

The great challenge ahead for the Merrimack region is to preserve its recent gains. The pressures of success are beginning to build. Now that the river no longer reeks, development on and near its banks is disrupting bird migration routes, and less dramatic sources of pollution, such as runoff of household waste, are growing in relative importance and proving far more difficult to control. The Merrimack has finally been restored to something like economic and environmental health. It would be a shame if its future were to be marked with the same shortsightedness that served it so poorly in the past.

Cait Murphy and Rosanne Haggerty's exploration of another nation-transforming waterway, the Erie Canal, ran in the April 2001 issue.

From *American Heritage*, April/May 2003, pp. 60-67.

Unscrambling the City

Archaic zoning laws lock cities into growth patterns that hardly anybody wants. Changing the rules can help set them free.

BY CHRISTOPHER SWOPE

Take a walk through Chicago's historic Lakeview neighborhood, and the new houses will jump right out at you. That's because they're jarringly incompatible with the old ones. On one quiet tree-lined street, you'll find a row of old two-story colonials with pitched roofs. Then you walk a little farther and it seems as though a giant rectangular box has fallen out of the sky. The new condominium building is twice as high as its older neighbors and literally casts shadows over their neat flower gardens and tiny front yards. Angry Lakeview residents have seen so many new buildings like this lately that they have come up with a sneering name for them. They call them "three-flats on steroids."

Listening to the complaints in Lakeview, you might wonder whether home builders are breaking the law and getting away with it, or at least bending the rules quite a bit. But that's not the case. If you take some time and study Chicago's zoning law, you'll find that these giant condos are technically by the book. It's not the new buildings that are the problem. The problem is Chicago's zoning ordinance. The code is nearly half a century old, and it is an outdated mishmash of vague and conflicting rules. Over the years, it has been amended repeatedly, to the point of nonsense. Above all, it's totally unpredictable. In Lakeview, zoning can yield anything from tasteful two-flats to garish McMansions, with no consideration at all for how they fit into the neighborhood.

Chicago's zoning problem lay dormant for decades while the city's economy sagged and population declined. Back in the 1970s and '80s, not much building was going on. But then the 1990s brought an economic boom and 112,000 new residents. While almost everyone is happy that the construction machine has been turned back on, so many Chicagoans are appalled by the way the new construction looks that Mayor Richard M. Daley decided it was time to rewrite the city's entire zoning code. Everything about Chicago land use is on the table: not just residential development but commercial and industrial as well. It is the largest overhaul of its kind in any U.S. city in 40 years.

But while few communities are going as far as Chicago, many are coming to a similar conclusion: The zoning laws on their books—most of them written in the 1950s and '60s—are all scrambled up. They are at once too vague and too complicated to produce the urban character most residents say they want.

The zoning problem afflicts both cities and suburbs and manifests itself in countless ways. It takes the form of oversized homes and farmland covered in cookie-cutter housing developments. It shows up as a sterile new strip mall opening up down the street from one that is dying. It becomes an obstacle when cities discover how hard it is to revive pedestrian life in their downtowns and neighborhood shopping districts. And it becomes a headache for city councils that spend half their time interpreting clumsy rules, issuing variances and haggling with developers.

What urban planners disagree about is whether the current system can be salvaged, or whether it should be scrapped altogether. Most cities are not ready to take the ultimate step. Chicago isn't going that far. Neither did Boston, Milwaukee, San Diego and San Jose. All of them retained the basic zoning conventions, even as they slogged through the process of streamlining the codes and rewriting them for the 21st century. According to researcher Stuart Meck, of the American Planning Association, there's a cyclical nature to all this. He points out that it's common for cities to update their laws after the sort of building boom many have enjoyed recently. "Cities are in growth mode again," Meck says, "but they're getting development based on standards that are 20, 30 or 40 years old."

MYRIAD CATEGORIES

For much of the past century, if you wanted to find out the latest thinking about zoning, Chicago was a good place to go. In 1923, it became one of the first cities, after New York, to adopt a zoning law. The motivation then was mostly health and safety. Smoke-spewing factories were encroaching on residential

PICTURE-BOOK ZONING

While Chicago and a few other large cities struggle to update old zoning laws for the new century, some places are going in a new direction. They are experimenting with zoning concepts percolating out of the New Urbanist movement, writing codes that bear a closer resemblance to picture books than to laws. Conventional zoning, they have decided, is based on an abstract language that leaves too much to chance. They would rather start with a question—what does the community want to look like—and then work back from there. "It's not enough to change the zoning," says New Urbanist author Peter Katz. "Cities have to move to a new system. They should look at the streets they like and the public spaces they like and then write the rules to get more of what they like and less of what they don't. Conventional zoning doesn't do that. It just gives a use and a density and then you hope for the best."

On jurisdiction currently buying in to this new idea is Arlington, Virginia, a suburb of 190,000 people just across the river from Washington D.C. A few months ago, Arlington's county board adopted a "form-based" zoning code for a 3.5-mile corridor known as Columbia Pike, making it one of the largest experiments yet with this new idea.

Columbia Pike is a typical traffic-choked suburban drag, lines mostly with strip malls, drive-throughs and apartment complexes ringed by parking lots. Developers have ignored the area for years. County planners want to convert it into a place that more closely resembles a classic American Main Street. They want a walkable commercial thoroughfare, featuring ground-floor retail blended together with offices and apartments above. But the old zoning code made this nearly impossible.

Rather than starting with a clear vision of what Arlington wants Columbia Pike to look like, the old code starts with a letter and a number: "C-2." The "C" stands for commercial uses only, and the "2" means that development should be of a medium density. C-2 is so vague that it could yield any number of building types. But the code's ambiguities don't end there. Building size is regulated by "floor area ratio," a calculation that again says nothing about whether the building should be suitable for a Main Street or an interstate highway exit. Finally, the code doesn't say where on the lot the building should go—just that it shouldn't sit near the roadway. Mostly, developers have used this recipe to build strip malls. "The code is really absolute on things that don't matter to us at all," says Arlington board member Chris Zimmerman. "The tools are all wrong for the job we're trying to do."

The new code for Columbia Pike abandons these old tools. It begins with a picture: What does a Main Street look like? Rather than abstract language, the new code uses visuals to show the form that the buildings should take. Buildings are three to six stories tall. And they sit on the sidewalk, with ground-floor windows and front doors, not 50 feet back from the street.

Compared with traditional zoning, a form-based code doesn't focus on specific uses. It specifies physical patterns. Whether the buildings are occupied by coffee shops, law offices or upstairs renters makes little difference. "Traditionally," says Peter Katz, "zoning stipulates a density and a use and it's anyone's guess whether you'll get what the planners' renderings look like. Form-based codes give a way to achieve what you see in the picture with precision."

One of the most prominent New Urbanists, Miami architect Andres Duany, advocates taking the form-based idea even further. In Duany's view, it's not only buildings along a road like Columbia Pike that should be coded according to physical form rather than use: entire metropolitan regions should be thought of this way. Duany is pushing an alternative he calls "Smart Code."

The Smart Code is based on the concept of the "transect." The idea is that there is a range of forms that the built environment can take. At one end is downtown, the urban core. At the other end is wilderness. In between are villages, suburbs and more dense urban neighborhoods. As Duany sees it, conventional zoning has failed to maintain the important distinctions between these types of places. Instead, it has made each of them resemble suburbia. When suburban building forms encroach on wilderness, the result is sprawl. When they encroach on urban areas, the result is lifeless downtowns.

Nashville-Davidson County, Tennessee, is one of the first places to begin incorporating these concepts into its planning process. The transect isn't a substitute for a zoning code, says planning director Rick Bernhardt. But it helps planners think about how one part of the city fits into the region, and how to zone accordingly. "It's really understanding what the purpose is of the part of the community you're designing," Bernhardt says, "and then making sure that the streetscape, the intensity and the mix of land use are all consistent with that."

—C.S.

neighborhoods, and the city's first ordinance sought to keep them out. By the 1950s, when more people drove cars, Chicago was a pioneer in rewriting the code to separate the places people live in from where they work and where they shop.

The 1957 zoning law was largely the creation of real estate developer Harry Chaddick, who proclaimed that the city was "being slowly strangled" by mixed uses of property. It classified every available parcel of land into myriad categories based on density. Residential neighborhoods, for example, were laid out in a range from "R1" (single family homes) to "R8" (high-rises). Land use rules were so strict as to dictate where ice cream shops, coin stores and haberdasheries could go. Chaddick's code was hailed in its time as a national model.

But over the years, one patch after another in the 1957 law made it almost impossible to use. Some parts contradicted other parts. Two attorneys could read it and come away with completely opposite views of what the code allowed. Finally, in 2000, the mayor tapped Ed Kus, a longtime city zoning attorney, to take charge of a full-scale rewrite. Kus thinks the law in the works will be equally as historic as Chaddick's—and more durable. "I hope the ordinance we come up with will be good for the next 50 years," Kus says.

Besides its rigidity, the old code has been plagued by false assumptions about population growth. Back in the 1950s, Chicago was a city of 3.6 million people, and planners expected it to reach a population of 5 million. Of course, it didn't work out that way. Like every other major city, Chicago lost a huge proportion of its residents to the suburbs. By 1990, it was down to fewer than 2.8 million residents. But it was still zoned to accommodate 5 million.

That's essentially how Lakeview got its three-flats on steroids. Had the city's population grown as the code anticipated, it would have needed a supply of large new residential buildings to replace its traditional two-flats and bungalows. The law made it possible to build these in lots of neighborhoods, regardless of the existing architecture or character.

For decades, this made relatively little difference, because the declining population limited demand for new housing in most of the city. Once the '90s boom hit, however, developers took advantage. They bought up old homes and tore them down, replacing them with massive condo projects. They built tall, and sometimes they built wide and deep, eating up front yards and side yards and often paving over the back for parking. "Developers are building to the max," Kus says. "We have all these new housing types and the zoning ordinance doesn't govern them very well."

There are other glaring problems. Although many people think of the 1950s as the decade when America went suburban, most retail business in Chicago was still conducted in storefronts along trolley lines, both in the city and the older close-in suburbs. The code reflects that mid-century reality. Some 700 miles of Chicago's arterial streets are zoned for commercial use, much more than the current local retail market can bear. Worse, the old code is full of anachronistic restrictions on what kinds of transactions can be conducted where. A store that sells computers needs a zoning variance to set up shop next door to one that fixes them. "If you're in a 'B1' district"—a neighborhood business corridor—"you can hardly do any business," Kus says.

All of these archaic provisions are quietly being reconsidered and revised on the ninth floor of city hall, where Kus heads a small team that includes two planning department staffers and a consultant from the planning firm of Duncan Associates. Their work will go to the zoning reform commission, a panel whose 17 members were picked by the mayor to hold exhaustive public meetings and then vote on the plan. The commission includes aldermen, architects, planners, business representatives and a labor leader. Developers are conspicuously absent, which may come back to bite the whole project later. But for now, the rewrite is moving remarkably fast. The city council is expected to pass the new code this fall. That will set the stage for an even more difficult task: drawing new maps to fit the changed rules.

In the past, Chicago's zoning reforms sought nothing less than to transform the face of the city. This time, however, there is more of a conservationist bent. What the reformers are trying to do is to lock in the qualities Chicagoans like about their oldest, most traditional neighborhoods. That's not to say they want to freeze the city in place. The building boom is quite popular. But it's also widely accepted that the character of Chicago's neighborhoods is the reason why the city is hot again, and that zoning should require new buildings to fit in. "Cities that will succeed in the future are the ones that maintain a unique character of place," says Alicia Mazur Berg, Chicago's planning commissioner. "People choose to live in many of our neighborhoods because they're attractive, they have front yards and buildings of the same scale."

MADE FOR WALKING

The new rules being drafted for residential areas are a good example of this thinking. Height limits will prevent new houses from towering over old ones. Neighborhoods such as Lakeview will likely be "downzoned" for less density. New homes will be required to have a green back yard, not a paved one, and builders will not be allowed to substitute a new creation known as a "patio pit" for a front yard. Garages will be expected to face an alley—not the street—and blank walls along the streetscape will be prohibited.

In the same spirit, the creators of the new zoning code are also proposing a new category, the Pedestrian Street, or "P-street." This is meant for a neighborhood shopping street that has survived in spite of the automobile and still thrives with pedestrian life. The new code aims to keep things that way. Zoning for P-streets will specifically outlaw strip malls, gas stations and drive-throughs, or any large curb cut that could interrupt the flow of pedestrians. It also will require new buildings to sit right on the sidewalk and have front doors and windows so that people walking by can see inside.

There are dozens of other ideas. The new code aims to liven up once-vibrant but now-dying neighborhood commercial streets by letting developers build housing there. For the first time ever, downtown Chicago will be treated as a distinct place, with its own special set of zoning rules. The code will largely ignore meaningless distinctions between businesses, such as whether they sell umbrellas or hats.

The new code also will recognize that the nature of manufacturing has changed. Light manufacturing will be allowed to mix with offices or nightclubs. But heavy industry will get zones of its own, not so much for the health reasons that were important in 1923 and 1957, but because the big manufacturers want it that way and Chicago doesn't want to lose them.

For all the changes, Chicago is still keeping most of the basic zoning conventions in place. It is also keeping much of the peculiar language of zoning—the designations such as "R2" and "C3" that sound more like droids from Star Wars than descriptions of places where people live, work and shop.

On the other hand, the new code will be different from the old code in one immediately identifiable way: It will be understandable. Pages of text are being slimmed down into charts and graphics, making the law easier to use for people without degrees in law or planning. An interactive version will go up on the city's Web site. "Predictability is important," says Ed Kus. "The average person should be able to pick up the zoning code and understand what can and can't be built in his neighborhood."

An Inner-City Renaissance

The nation's ghettos are making surprising strides. Will the gains last?

Take a stroll around Harlem these days, and you'll find plenty of the broken windows and rundown buildings that typify America's ghettos. But you'll also see a neighborhood blooming with signs of economic vitality. New restaurants have opened on the main drag, 125th Street, not far from a huge Pathmark supermarket, one of the first chains to offer an alternative to overpriced bodegas when it moved in four years ago. There's a Starbucks—and nearby, Harlem U.S.A., a swank complex that opened in 2001 with a nine-screen Magic Johnson Theatres, plus Disney and Old Navy stores and other retail outlets. Despite the aftermath of September 11 and a sluggish economy, condos are still going up and brownstones are being renovated as the middle classes—mostly minorities but also whites—snap up houses that are cheap by Manhattan standards.

It's not just Harlem, either. Across the U.S., an astonishing economic trend got under way in the 1990s. After half a century of relentless decline, many of America's blighted inner cities have begun to improve. On a wide range of economic measures, ghettos and their surrounding neighborhoods actually outpaced the U.S. as a whole, according to a new study of the 100 largest inner cities by Boston's Initiative for a Competitive Inner City, a group founded in 1994 by Harvard University management professor Michael E. Porter.

Consider this: Median inner-city household incomes grew by 20% between 1990 and 2000, to a surprising $35,000 a year, the ICIC found, while the national median gained only 14%, to about $57,000. Inner-city poverty fell faster than poverty did in the U.S. as a whole, housing units and homeownership grew more quickly, and even the share of the population with high school degrees increased more. Employment growth didn't outdo the national average, with jobs climbing 1% a year between 1995 and 2001, vs. 2% nationally. Still, the fact that inner cities, which are 82% minority, created any jobs at all after decades of steady shrinkage is something of a miracle.

SCENT OF OPPORTUNITY

NOR ARE THE GAINS just the byproduct of the superheated economy of the late 1990s. Rather, they represent a fundamental shift in the economics of the inner city as falling crime rates and crowded suburbs lure the middle-class back to America's downtowns. After decades of flight out of inner cities, companies as diverse as Bank of America, Merrill Lynch, and Home Depot have begun to see them as juicy investment opportunities. National chains are opening stores, auto dealerships, and banks to tap into the unfulfilled demand of inner cities.

Wall Street, too, is jumping in, making loans and putting up equity for local entrepreneurs. "Smart businesspeople gravitate toward good opportunities, and it has become clear that inner cities are just that," says David W. Tralka, chairman of Merrill Lynch & Co.'s Business Financial Services group. In 2002, his group, which caters to small business, began formally targeting inner cities. It now offers financing and commercial mortgages for hundreds of inner-city entrepreneurs around the country.

Is it possible that America at last has started to solve one of its most intractable social ills? True, the progress so far is minuscule compared with the problems created by decades of capital flight, abysmal schools, and drug abuse. And some inner cities, like Detroit's, have made little sustained progress. Ghettos also have been hit by the joblessness of this latest recovery. The national poverty rate has jumped by nearly a percentage point since 2000, to 12.1% last year, so it almost certainly did likewise in inner cities, which the ICIC defined as census tracts with poverty rates of 20% or more.

But as the economy recovers, a confluence of long-term trends is likely to continue to lift inner cities for years. The falling crime rate across the country has been a key factor, easing fears that you take your life in your hands by setting foot in an inner city. At the same time, larger demographic shifts—aging boomers turned empty nesters, more gays and nontraditional households without children, homeowners fed up with long commutes—have propelled Americans back into cities. When they arrive, slums suddenly look like choice real estate at bargain prices.

BEYOND PHILANTHROPY

POLITICAL AND CIVIC LEADERS helped lay the groundwork, too. After floundering for decades following the exodus of factories to the suburbs in the 1950s, many cities finally found new economic missions in the 1990s, such as tourism, entertainment, finance, and services. This has helped boost the geographic desirability of inner-city areas. New state and federal policies brought private capital back, too, by putting teeth into anti-redlining laws and by switching housing subsidies from public projects to tax breaks for builders. As a result, neighborhoods like the predominantly African-American Leimert Park in South Central Los Angeles are becoming thriving enclaves.

The outcome has been a burst of corporate and entrepreneurial activity that already has done more to transform inner

Inner Cities and Their Residents ...

The Boston-based Initiative for a Competitive Inner City has completed the first-ever analysis of the 100-largest inner cities in the U.S. and finds the once-dismal picture brightening

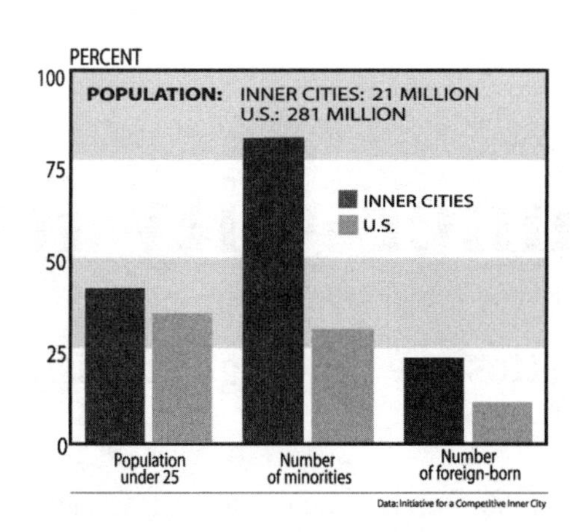

...Did Better than the Nation As a Whole in the 90s...

Change between 1990 and 2000 Census data

	INNER CITIES	U.S.
Population	24%	13%
Household income	20%	14%
Housing unit growth	20%	13%
High school graduates*	55 to 61%	75 to 80%
College graduates*	10 to 13%	20 to 24%
Home ownership	29 to 32%	64 to 66%
Poverty rate	34 to 30%	13 to 12%

*Of those 25 and over

...Although There's Still A Long Way to Go

2000 Census data

	INNER CITIES	U.S.
High school graduates*	61%	80%
College graduates*	13%	24%
Poverty rate	31%	11.3%
Unemployment rate	12.8%	5.8%
Home ownership rate	32%	66%
Average household income	$34,755	$56,600
Aggregate household income	$250 billion	$6 trillion

Data: Initiative for a Competitive Inner City

cities than have decades of philanthropy and government programs. "What we couldn't get people to do on a social basis they're willing to do on an economic basis," says Albert B. Ratner, co-chairman of Forest City Enterprises Inc., a $5 billion real estate investment company that has invested in dozens of inner-city projects across the country.

EMERGING MARKETS

THE NEW VIEW OF GHETTOS began to take hold in the mid-1990s, when people such as Bill Clinton and Jesse Jackson started likening them to emerging markets overseas. Porter set up the ICIC in 1994 as an advocacy group to promote inner cities as overlooked investment opportunities. Since then, it has worked with a range of companies, including BofA, Merrill Lynch, Boston Consulting Group, and PricewaterhouseCoopers to analyze just how much spending power exists in inner cities.

The new study, due to be released on Oct. 16, uses detailed census tract data to paint the first comprehensive economic and demographic portrait of the 21 million people who live in the 100 largest inner cities. The goal, says Porter, "is to get market forces to bring inner cities up to surrounding levels."

Taken together, the data show an extraordinary renaissance under way in places long ago written off as lost causes. America's ghettos first began to form early in the last century, as blacks left Southern farms for factory jobs in Northern cities. By World War II, most major cities had areas that were up to 80% black, according to the 1993 book *American Apartheid*, co-authored by University of Pennsylvania sociology professor Douglas S. Massey and Nancy A. Denton, a sociology professor at the State University of New York at Albany. Ghettos grew faster after World War II as most blacks and Hispanics who could follow manufacturing jobs to the suburbs did so, leaving behind the poorest and most un-employable. Immigrants poured in, too, although most tended to leave as they assimilated.

In this context, the solid gains the ICIC found in the 1990s represent an extraordinary shift in fortunes. One of the biggest changes has come in housing. As cities have become desirable places to live again, the number of inner-city housing units jumped by 20% in the 1990s, vs. 13% average for the U.S. as a whole.

A number of companies were quick to see the change. BofA, for example, has developed a thriving inner-city business since it first began to see ghettos as a growth market six years ago. In 1999 it pulled together a new unit called Community Development Banking, which focuses primarily on affordable housing for urban, mostly inner-city, markets, says CDB President Douglas B. Woodruff. His group's 300 associates are on track this year to make $1.5 billion in housing loans in 38 cities, from Baltimore to St. Louis.

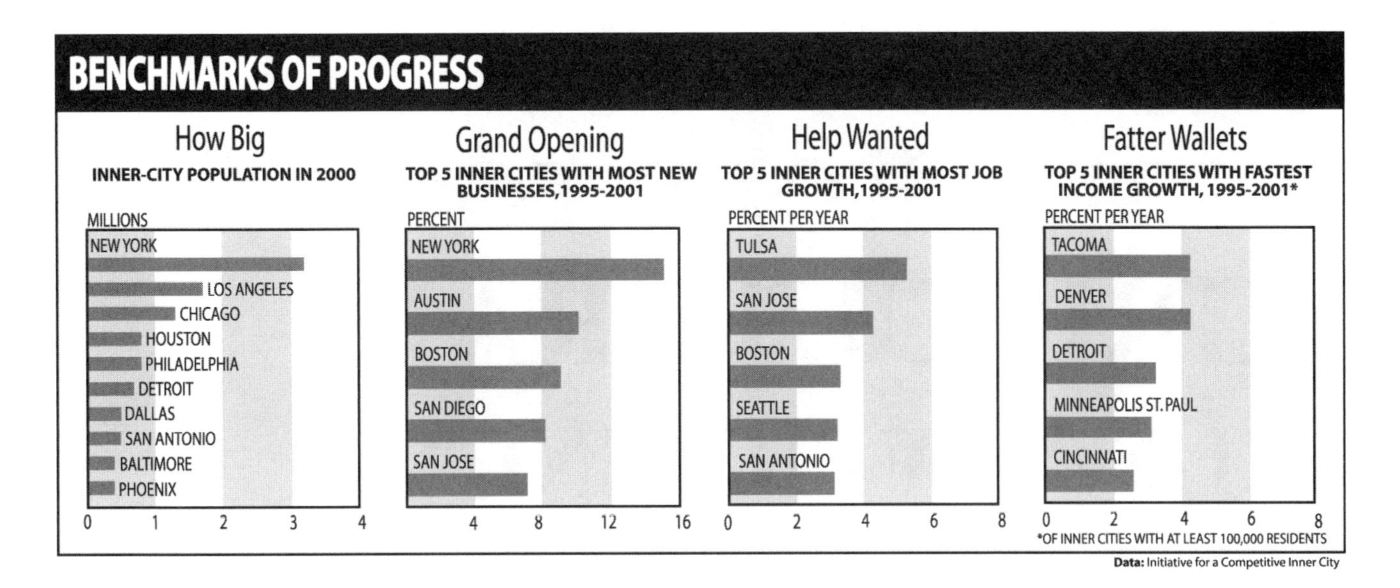

They will do an additional $550 million in equity investments, mostly real estate.

SHELLED OUT

PENSION FUNDS AND other large investors are putting in cash, too. The Los Angeles County Employee Retirement Assn. has sunk $210 million into urban real estate since 2000, including $87 million in August for a bankrupt, 2,496-room apartment complex in Brooklyn, N.Y. The plan is to do things like fix the broken elevators, hire security guards, and kick out nonpaying tenants. "We believe there are opportunities that weren't there before or that we weren't aware of," says board member Bruce Perelman.

One question is whether the ICIC's findings represent not so much progress by the poor as their displacement by middle-class newcomers. In other words, inner-city incomes could be rising simply because affluent new home buyers jack up the average. But experts think gentrification explains only a small part of what's going on. "It's certainly a local phenomenon, but if you aggregate 100 inner cities, gentrification is a small trend," says ICIC research director Alvaro Lima, who spearheaded the study.

In Chicago, for instance, a $65 million redevelopment of the notorious Cabrini-Green housing project has replaced three slummy high-rises with mixed-income units. The area has a new library, new schools, and a new retail center featuring a major grocery store, Starbucks, and Blockbuster (BBI)—all staffed by scores of local residents. "The goal is not gentrification, it's to integrate the classes," says Phyllis L. Martin, the head of a local committee that's trying to lure more than $50 million in private capital to help the city replace 3,245 public housing units in another blighted area, Bronzeville.

"We're just beginning to undo all the damage"

Despite the brightening picture, the decay of most inner cities is so advanced that half a dozen years of progress makes only a dent. The degree of poverty—a measure of how many poor people there are in a census tract—fell 11% in 60 large cities in the 1990s, according to an analysis by U Penn's Massey that parallels ICIC's approach. While that's a significant decline, it only begins to offset the doubling of poverty concentrations in prior decades, he found. "The gains are the first positive news since at least the 1950s, but we're just beginning to undo all the damage," says Massey.

BADGE OF SHAME

WHAT'S MORE, too many inner cities remain untouched. More than a third of the ICIC's 100 cities lost jobs between 1995 and 2001. Detroit's ghetto has seen little new development and shed one-fifth of its jobs over this period. Residents did gain from the booming auto industry, which hired many locals and pushed up their median incomes at a 3.2% annual pace in the 1990s—the third highest increase of the ICIC 100. But with auto makers now shedding jobs again, those gains are likely to be short-lived. More broadly, improving inner cities won't come close to wiping out poverty in the U.S. While the inner-city poverty rate of 31% is nearly three times the national average, the 6.5 million poor people who live there represent less than a fifth of the country's 34.6 million poor.

Still, America's ghettos have been a national badge of shame for so long that any real gain is news. The change in perspective also seems to be an enduring one, not just a 1990s blip. For evidence, consider Potamkin Auto Group, which owns 70 dealerships around the country and will break ground in Harlem in late October on a $50 million development that will include Cadillac, Chevrolet, Hummer, and Saturn dealers. Potamkin also has a project in another inner city and is mulling a national expansion. "We see opportunities there," says Robert Potamkin, president of the family-owned company. This view, that inner cities can be a good place to do business, may be the most hopeful news about the country's urban blight in decades.

—By Aaron Bernstein in Washington, with Christopher Palmeri in Los Angeles and Roger O. Crockett in Chicago

ON THE ROAD TO AGRICULTURAL SELF-SUFFICIENCY

As it has in many other areas of human endeavor, Saudi Arabia has achieved spectacular success in the transformation of its agricultural sector over the past three decades. During this period, large areas of the desert have been transformed into productive agricultural fields, bringing forth an abundance of fruits, vegetables and grains, and making a nation that had been largely dependent on the import of much of its foodstuffs into one that is now self-sufficient in poultry, fruits, vegetables and grains.

Saudi Arabia's agricultural sector has made the country self-sufficient in the production of major food products, and an exporter of fruit, vegetables and flowers.

The transformation of agriculture in Saudi Arabia is the more astounding because of the severe limitations that nature has placed on the land. For a country most of whose landmass is covered with vast deserts, including the *Rub' Al-Khali,* the largest sand desert in the world, and which lacks permanent rivers and receives only an average of about four inches of rain annually, one of the lowest rates in the world, aspiring to agricultural self-sufficiency is a major challenge.

Historically, outside of a small coastal strip in the southwestern part of what is now Saudi Arabia, agricultural activity in the arid Arabian peninsula was limited to date farming and small-scale vegetable production, and then only in the widely scattered oases. The valuable water percolating to the surface in springs and pumped up by draught animals from bore wells at these sites was traditionally distributed in the oases through a network of fields, where shaded by date palms, small plots of fertile land were planted with cereals, grains and vegetables, producing enough food for the local communities, with any excess sold to passing caravans.

The limited water supply dictated the small scale of agriculture, and therefore was a chief factor in keeping a check on the peninsula's population, which at the start of the 20th century was largely the same as it had been for generations.

The people of the peninsula were exposed early in human history to advances in agriculture that took place in the Fertile Crescent, an arc of land stretching from Mesopotamia through southern Anatolia into the eastern Mediterranean littoral. It was in this area that man made the transition from hunter-gathering to living in settled communities supported by agriculture. Irrigation, domestication of farm animals and other innovations were also first introduced in this area. While these advances had a profound impact on life in the Fertile Crescent, where large river systems provided vast quantities of water throughout the year, resulting in massive population explosions, their effect on the people of the peninsula, where rivers ran only after the sporadic rainfalls of the winter season and were dry for the rest of the year, were minimal due to limited water supplies.

Using innovative irrigation techniques, such as center-pivot systems, Saudi farmers turn the desert from an arid landscape to productive fields.

Shortly after unifying the peoples of the larger part of the peninsula into the modern Kingdom of Saudi Arabia in 1932, King Abdulaziz bin Abdulrahman Al-Saud undertook the first steps of an ongoing effort to expand agriculture. The security and stability that his rule brought to the country encouraged greater trade between various parts of the Kingdom, with agricultural products accounting for a sizeable portion of that exchange.

In the following decades, the Kingdom embarked on a sustained effort to bring about a fundamental change in the character of agriculture, away from traditional, small-scale farming, to larger-scale modern methods. In the 1940s, the government actively encouraged villagers and farmers to increase the land under cultivation by exempting from import duties agricultural machinery, including pumps for wells and irrigation, and later providing interest-free loans and financial assistance for their purchase.

In 1948, the Directorate of Agriculture was established to bring all aspects of agriculture under the aegis of a central authority empowered to take all practical steps to expand the sector. The new office was given the task of locating, mapping, regulating and improving the country's water resources. It soon began providing water pumps and other irrigation machinery free of charge to farmers, building dams in the *wadis* [dry riverbeds] to collect the runoff from flash floods that previously ran off into the desert, providing arable land to farmers and extending loans to help them get started. The directorate was expanded and transformed into the Ministry of Agriculture and Water in 1953, allowing it to play a more active and efficient role in the ongoing development of the agricultural sector.

Modern agro-industrial ventures breed dairy cows, produce fodder and operate computerized milking stalls to provide enough dairy products to meet domestic demand and export to nearby markets.

With the introduction of the First Five-Year Development Plan in 1970, Saudi Arabian agriculture entered a new era of rapid growth that has made possible today's accomplishments. The development of the agricultural sector was meant not only as a way of reducing reliance on imports and achieving self-sufficiency in food production, but also as a means of diversifying the economy away from oil. Saudi Arabia undertook to expand agricultural output in the traditionally fertile areas as well as extend it to other regions that have fertile soil but lack sufficient water. To provide the water needed to irrigate new agricultural fields, the Kingdom took steps to collect and conserve vast quantities of water from the frequent flash floods that occur in certain parts of the country. To collect the runoff, the Ministry of Agriculture and Water built more than 200 dams placed strategically in *wadis* that have a total reservoir capacity of more than 16 billion cubic feet of water. It has also established a vast network of irrigation canals to distribute water from these reservoirs to nearby farms. Additionally, water from underground reservoirs, accessed via deep wells, has been used to meet agricultural needs. Another source of water that is growing every year is treated urban wastewater, which is used exclusively for irrigating agricultural crops.

As it has in other areas of economic and social development, the government has applied its vast resources to a systematic, long-term program designed to achieve sustainable agricultural growth. As examples, the government realized that a modern transportation network is essential to the economic health of the agricultural sector. Therefore, it moved to build tens of thousands of miles of rural roads needed to transport machinery to farms and bring their products to market in a timely fashion. It also built a network of storage and export facilities to allow the shipment of excess Saudi agricultural products to markets outside the Kingdom; established a large number of agricultural colleges; and helped fund and set up training institutions.

At the same time, the government introduced the land distribution and reclamation program as a means of distributing fallow land free of charge, mostly in small- and medium-sized plots, as a means of increasing the area under cultivation and encouraging crop and livestock production. The beneficiaries are required to develop a minimum of one quarter of the land surface within two to five years. Upon compliance, full ownership of the land is transferred to the farmer. The government continues to assist new farmers in implementing capital-intensive projects with special emphasis on diversification and greater efficiency.

The government also established the Grain Silos and Flour Mills Organization in 1972 to purchase and store wheat, construct flour mills and produce animal feed to support the nationwide growth of agriculture. The organization now purchases millions of tons of domestic grain annually.

To raise farm productivity, the government also funds and supports research projects aimed at producing new food crops to increase harvest and develop plant strains with greater resistance to pests. These programs are conducted in cooperation between local farmers and scientists at agricultural research facilities at Saudi Arabian universities and colleges.

These efforts collectively have helped transform vast tracts of the desert into fertile farmland. Land under cultivation grew from under 400,000 acres in 1976 to some 11 million acres today.

Combines harvest wheat from fields wrested from the desert, making the Kingdom a net exporter of grain products. Saudi Arabia is also self-sufficient in the production of fodder and other agricultural products that previously had to be imported at great cost.

Wheat is one example of the Kingdom's dramatic agricultural accomplishments that were largely achieved by the private sector. Saudi Arabia had traditionally produced several thousand tons of wheat annually but was a major importer of the grain. Beginning in the 1970s, the government encouraged farmers to increase wheat production to achieve self-sufficiency in this vital component of the daily food intake. Subsidies, loans and technical assistance provided by the government brought about a steady increase in the land under wheat cultivation, as well as a parallel gain in yields per unit of land, reaching 3.6 tons per acre in some areas. By 1984, the Kingdom had become self-sufficient in wheat, and by the early 1990s was producing some four million tons annually, four times the national demand. Shortly thereafter, Saudi Arabia began selling wheat to some thirty countries, while supplying large shipments as humanitarian assistance to countries to suffering from famine and drought. A similar policy was adopted for the production of barley, sorghum and millet, with identical results. However, since production exceeded domestic need, and it was realized that the water needed to grow the excess supplies of these grains was depleting the nation's underground aquifers, in 1996 the government suspended subsidies and thereby reduced output of wheat to 1.1 million tons and barley to 460,000 tons, levels adequate to meet domestic demand.

Almost all agricultural activities are fully mechanized, relying on machinery and state-of-the-art techniques.

Perhaps more than any other sector of the economy, the development of Saudi agriculture owes its success to the private sector. Unlike industry, where the government took the lead to establish factories and manufacturing facilities, then encouraged the private sector to follow, Saudi individuals and businesses were the leaders in the growth of agriculture, albeit with extensive government support and assistance.

To encourage the private sector to set the pace, the government introduced an array of programs, including the provision of long-term, interest-free loans and technical and support services beginning in the 1970s. It also provided other incentives, including the provision of free seeds and fertilizers, low-cost water, fuel and electricity, and duty-free imports of raw materials and machinery. To encourage foreign investments in agriculture, the government has exempted foreign joint-venture partners of Saudi individuals or companies from paying taxes for a period of up to 10 years, and the new investment regulations of April 2000 offer further incentives.

Saudi dairy industries, such as the Al-Safi Dairy Company in Al-Kharj, produce milk, yoghurt, cheeses and other products in fully-automated plants.

While the primary agency responsible for implementing agricultural policy over the past half century has been the Ministry of Agriculture and Water, which was separated into the Ministry of Agriculture and the Ministry of Water in 2002, the Saudi Arabian Agricultural Bank (SAAB) has also played a vital role in bringing about the country's phenomenal agricultural growth by disbursing subsidies and granting interest-free loans. Over the past three decades, the bank has provided tens of billions of dollars of such grants and loans, enabling individual farmers and agricultural companies to expand their activities.

As a result, Saudi farmers and businesses have brought about a complete transformation of the agricultural sector and helped the kingdom achieve self-sufficiency in many fields.

One example is the Al-Safi Dairy Company in Al-Kharj, 60 miles southeast of Riyadh, which according to the Guinness Book of World Records is the largest dairy operation in the world. Owned by Saudi businessmen, Al-Safi started operation in 1979 by importing 3,500 Holstein Friesian dairy cattle and establishing a 25,000-acre farm. By 1981, the company was producing 10,000 gallons of milk daily for domestic consumption. A completely self-contained enterprise that breeds cattle, produces feed and operates its own processing and packaging factories, the company has grown substantially over the past two decades. Today, Al-Safi produces some 100,000 gallons of milk per day from its herd of 25,000 dairy cattle. The operation relies on computers to keep tabs on all aspects of a state-of-the-art venture, from breeding cows and monitoring their health and production levels to growing all the feed needed by the herd. Al-Safi's dairy cows have some of the highest milk output levels in the world, and its daily output is pasteurized, processed and packaged in modern factories for distribution as milk, cheese, yoghurt and other products.

Joint venture partnerships between Saudi and foreign companies in the agricultural and food-processing industries include the United Sugar Company plant in Jeddah which produces enough sugar and molasses for consumers and industry.

United Sugar Company (USC) is another example of the phenomenal success of the agriculture and food production industry in Saudi Arabia. Saudi consumers and industries require some 530,000 tons of sugar annually, all imported from abroad until the late 1990s. In 1997, USC inaugurated in Jeddah what is described as the most up-to-date sugar refinery and the fifth largest such operation in the world. By the end of that year the company was meeting almost all domestic demand, with one-third of its output in the form of fine sugar for consumer use and the remainder comprised of coarse sugar and molasses for use by food processing industries. The company is now in the process of expansion to meet future growth in domestic demand and also to export its products to other countries.

The Savola Company is one of the largest food manufacturers and processors in the Kingdom. A diversified company,

Savola has a large number of plants that produce a wide range of products. The chocolate, candy bars, pastries, potato chips and corn snacks produced by its Snack Foods Company account for 50 percent of all snack foods consumed in the Kingdom, and it exports to other Arab countries. The Savola Edible Oil Company manufactures, markets and distributes more than 15 different types of products, including corn, sunflower, palm and blended oils. In addition to plants in the Kingdom, the company also owns and operates edible oil facilities in Egypt and Bahrain, with its Al-Arabi and Afia brands being the most famous, selling more than 150,000 tons a year in Saudi Arabia and 100,000 tons in Egypt and Bahrain annually. Additionally, Savola's agro-industrial division produces tens of thousands of tons of potatoes, animal feed, and shrimps and fish at various facilities throughout Saudi Arabia. The Savola Packaging Systems Company manufactures all the glass, plastic, metal and cardboard canisters, bottles, cans and boxes that are used by the group's various factories, as well as other food processing facilities in the Kingdom.

> Using greenhouses, Saudi flower growers produce a variety of plants for the local market and export surplus as far away as Europe.

There are now hundreds of successful operations such as these on different scales throughout the Kingdom. Together they account for a phenomenal growth in agricultural production in the past three decades, with the value of output increasing more than 30-fold to over 17 billion U.S. dollars annually. Agriculture's share of the Kingdom's gross domestic product (GDP) has climbed from just 1.3 percent in 1970 to well over 10 percent.

Today, the Kingdom is a major producer of agricultural goods, and whereby previously output was limited to a few crops, the sector now enjoys great diversity. The Ministry of Agriculture reports that the Kingdom is producing nearly nine million tons of products annually. Saudi farmers produce some 2.2 million tons of wheat, barley, sorghum and millet, and nearly three million tons of alfalfa and other types of fodder for the livestock industry.

The Kingdom also produces 1.2 million tons of fruits annually, and is a major exporter to its neighbors. Among its most productive crops are watermelon, grapes and citrus fruits. This growth has been achieved by improving both agricultural techniques and developing strains that are best suited to the Kingdom's soil and climate. At Jizan in the country's well-watered southwest, the Al-Hikmah Research Station is producing tropical fruits including pineapples, paw-paws, bananas, mangoes and guavas that can be easily grown in Saudi Arabia.

> Saudi Arabia has preserved its traditional agriculture, including date farming, while venturing into new areas using new technology and machinery to produce new crops, such as soybeans.

Output of vegetables now stands at some 2.7 million tons. As a result, locally produced vegetables satisfy 84 percent of domestic consumption, whereas most of the vegetables consumed in the Kingdom were imported a few decades ago. Also, Saudi farmers are now exporting potatoes, carrots, tomatoes, onions, squash and pumpkins, with 60,000 tons shipped to neighboring countries every year.

Intensive dairy, meat, poultry and egg farming by small farmers and large-scale agro-industrial operations have satisfied local demand. These ventures produce 500,000 tons of poultry meat, 2.5 billion eggs, over 800,000 tons of dairy products and 160,000 tons of red meat every year.

While fish production through traditional off-shore fishing has been constantly on the increase, and now totals 60,000 tons a year, the Kingdom is exploring ways of further increasing its catch and encouraging greater private investment. One of the new areas in which the private sector is investing with government support is aquaculture. The number of fish farms, either using pens in the sea or tanks offshore, has been increasing steadily. Most are located along Saudi Arabia's Red Sea coast. Shrimp farming has been particularly successful. The Red Sea Shrimp Farm, for example, which is managed by Saudi hydrobiologists, marine engineers and aquaculture technicians produces more than 1,500 tons annually. Its shrimp, including the preferred black tiger, is exported mainly to the United States and to Japan.

As the past three decades have shown, Saudi agriculture is diversifying in new directions in a competitive market and expanding to meet the needs of a growing nation.

UNIT 4

Spatial Interaction and Mapping

Unit Selections

Key Points to Consider

- Describe the spatial form of the place in which you live. Do you live in a rural area, a town, or a city, and why was that particular location chosen?
- How does your hometown interact with its surrounding region? With other places in the state? With other states? With other places in the world?
- How are places "brought closer together" when transportation systems are improved?
- Discuss the importance of the geographical concept of accessibility.
- What problems occur when transportation systems are overloaded?
- How will public transportation be different in the future? Will there be more or fewer private autos in the next 25 years? Defend your answer.
- How good a map-reader are you? Why are maps useful in studying a place?

Links: www.dushkin.com/online/
These sites are annotated in the World Wide Web pages.

Edinburgh Geographical Information Systems
http://www.geo.ed.ac.uk/home/gishome.html

Geography for GIS
http://www.ncgia.ucsb.edu/cctp/units/geog_for_GIS/GC_index.html

GIS Frequently Asked Questions and General Information
http://www.census.gov/geo/www/faq-index.html

International Map Trade Association
http://www.maptrade.org

PSC Publications
http://www.psc.isr.umich.edu

Geography is the study not only of places in their own right but also of the ways in which places interact. Highways, airline routes, telecommunication systems, and even thoughts connect places. These forms of spatial interaction are an important part of the work of geographers.

"Mapping Opportunities," discusses the importance of GIS within the context of geographical concepts. Four articles on GIS follow. The first discusses geospatial asset management; the second deals with the gap between GIS technology and remote sensing data acquisition; the third covers public input to airport expansion through a GIS Website; and the fourth deals with the Oak Ridge National Laboratory and its role in the historical development of GIS products. "Mapping the Diversity of Nature" reviews an important research project in Middle America. Naval maps produced during World War II are reviewed in the next selection. "Resegregation's Aftermath" includes a map indicating failures in the K-12 education system. Managua, Nicaragua, devastated by an enormous earthquake in 1972, is a city literally without a map.

It is essential that geographers be able to describe the detailed spatial patterns of the world. Neither photographs nor words could do the job adequately, because they literally capture too much of the detail of a place. Therefore, maps seem to be the best way to present many of the topics analyzed in geography. Maps and geography go hand in hand. Although maps are used in other disciplines, their association with geography is the most highly developed.

A map is a graphic that presents a generalized and scaled-down view of particular occurrences or themes in an area. If a picture is worth a thousand words, then a map is worth a thousand pictures—or more! There is simply no better way to "view" a portion of Earth's surface or an associated pattern than with a map.

Mapping opportunities

Scientists who can combine geographic information systems with satellite data are in demand in a variety of disciplines. Virginia Gewin gets her bearings.

VIRGINIA GEWIN

Forest fires ravaging southern California, foot-and-mouth disease devastating the British livestock industry, the recent outbreak of severe acute respiratory syndrome (SARS)—all of these disasters have at least one thing in common: the role played by geospatial analysts, mining satellite images for information to help authorities make crucial decisions. By combining layers of spatially referenced data called geographic information systems (GIS) with remotely sensed aerial or satellite images, these high-tech geographers have turned computer mapping into a powerful decision-making tool.

Natural-resource managers aren't the only ones to take notice. From military planning to real estate, geospatial technologies have changed the face of geography and broadened job prospects across public and private sectors.

Earlier this year, the US Department of Labor identified geotechnology as one of the three most important emerging and evolving fields, along with nanotechnology and biotechnology. Job opportunities are growing and diversifying as geospatial technologies prove their value in ever more areas.

The demand for geospatial skills is growing worldwide, but the job prospects reflect a country's geography, mapping history and even political agenda. In the United States, the focus on homeland security has been one of many factors driving the job market. Another is its vast, unmapped landscape. While European countries are integrating GIS into government decision-making, their well-charted lands give them little need for expensive satellite imagery.

AN EXPANDING MARKET

All indications are that the US$5-billion worldwide geospatial market will grow to $30 billion by 2005—a dramatic increase that is sure to create new jobs, according to Emily DeRocco, assistant secretary at the US Department of Labor's employment and training division. NASA says that 26% of its most highly trained geotech staff are due to retire in the next decade, and the National Imagery and Mapping Agency is expected to need 7,000 people trained in GIS in the next three years.

Of the 140,000 organizations globally that use GIS, most are government agencies—local, national and international. A ten-year industry forecast put together last year by the American Society for Photogrammetry & Remote Sensing (ASPRS) identified environmental, civil government, defence and security, and transportation as the most active market segments.

Business at the Earth-imagery provider Space Imaging, of Thornton, Colorado, increased by 70% last year, says Gene Colabatistto, executive vice-president of the company's consulting service. To keep up momentum, the company plans to hire more recruits with a combination of technical and business skills. Colabatistto cites the increased adoption of GIS technologies by governments as a reason for the rise. He adds that the US military, the first industry to adopt GIS and remote sensing on a large scale, has spent more than $1 billion on commercial remote sensing and GIS in the past two years.

LOOKING DOWN IS LOOKING UP

The private sector hasn't traditionally offered many jobs for geographers, but location-based services and mapping—or 'geographic management systems'—are changing the field. "The business of looking down is looking up," says Thomas Lillesand, director of the University of Wisconsin's Environmental Remote Sensing Center in Madison, Wisconsin.

Imagery providers such as Digital Globe of Longmont, Colorado, also need more GIS-trained workers as markets continue to emerge. Spokesman Chuck Herring says that the company has identified 54 markets in which spatial data are starting to play a role.

The Environmental Systems Research Institute (ESRI), in Redlands, California, sets the industry standards for geospatial software. Most of its 2,500 employees have undergraduate training in geography or information technology, although PhDs are sought after to fill the software-development positions. Many private companies, including the ESRI and Space Imaging, offer valuable work experience to both graduate and final-year undergraduate students.

Graduates in natural-resource management note that GIS and remote-sensing skills are becoming as important as fieldwork. GIS platforms, which manipulate all forms of image data, are transforming disciplines such as ecology, marine biology and forestry.

"Science has discovered geography," says Doug Richardson, executive director of the Association of American Geographers (AAG). Many of the National Science Foundation's multidisciplinary research programmes now include a geospatial component.

SKILLED LABOUR

Some universities are offering two-year non-thesis master's programmes in geospatial technologies, including communication and business courses—perfect for professionals who want to build on existing skills or move into a new field. The non-profit Sloan Foundation has funded several geospatially related professional master's programmes. In addition, numerous short courses are available to bring professionals up to speed. Indeed, the ESRI alone trains over 200,000 people a year. AAG and ASPRS conferences also offer training sessions.

Although technical skills are important, Richardson stresses that employees need a deep understanding of underlying geographic concepts. "It's a mistake to think that these technologies require only technician-oriented functions," he says.

Throughout the European Union (EU), the many top-quality graduate geography programmes remain the primary training grounds. Recently, a few pan-European projects have also emerged, including a new international institute designed to train future geographers. Building on a collaboration between the European Space Agency and the US National Science Foundation, the Vespucci Initiative in 2002 began three-week summer workshops training students from around the world in spatial data infrastructure, spatial analysis and geodemographics. The EU even promotes distance learning: UNIGIS, a network of European universities, prides itself on being the only virtual, global, multilingual GIS programme in the world.

Although GIS is increasingly incorporated into UK government practices, there is little demand for remote-sensing expertise in this small and heavily mapped country. Mark Linehan, director of the London-based Association for Geographic Information, says that although the public-sector market is growing, it remains a struggle to find jobs for MScs at the appropriate pay scale and qualification level.

The European Commission (EC) is laying the groundwork to ease data-sharing across countries in anticipation of wider adoption of GIS among the member-state governments and to cut the costs of data gathering. That process alone will require at least a couple of thousand people trained in GIS, and many more proposals are expected.

Indeed, the EC and the European Space Agency have joined to propose a Global Monitoring for Environment and Security initiative, to provide permanent access to information on environmental management, risk surveillance and civil security. Given the scope of the mandate, this is likely to need people who understand how to interpret, integrate and manage satellite information—those who also have a background in natural-resource issues will be in highest demand.

Considering the role that GIS played in staving the spread of foot-and-mouth disease, such a system will not only increase the prevalence of geospatial skills in Europe, it will better connect data with Europe's resource managers.

Virginia Gewin is a freelance science writer in Corvallis, Oregon.

Geospatial Asset Management Solutions

How do we maintain and repair the nation's infrastructure, without going broke?

Damon D. Judd

Typical asset-intensive organizations that have an extensive infrastructure to build, maintain, repair, replace and ultimately decommission, require a substantial amount of work to be managed. In most cases they have or will invest in some type of asset and/or work management system. This is generally a database application that allows facilities, equipment, vehicles, and materials to be tracked and to enable work to be assigned to work crews, managed by their supervisors, and reported to upper management in some fashion.

There are several business drivers that dictate when work needs to be assigned and completed to maintain, repair, and operate the assets owned (or managed) by the organization. Some of those drivers include homeland security concerns, prevention of massive infrastructure failures such as the recent blackout in the northeast U.S., adherence to GASB 34 standards for governmental accounting practices, the effects of deregulation on the utility industry, and compliance with various environmental laws and regulations.

What is GIS Integration with Asset Management?

By integrating the capabilities of a Geographic Information System (GIS), which many organizations already have in place, with an Asset Management System (AMS), public works departments, investor-owned utilities, and other asset-intensive organizations can improve their ability to manage their asset inventories, and do so even more efficiently than ever before. The GIS enables map-based views of the asset and work information that is managed using an AMS.

Because there is a database relationship between the spatial data in the GIS and details of the assets and the work performed on those assets in the AMS, extended graphical display and data analysis functionality becomes possible. In an integrated solution, the graphical perspective can be presented along with the current condition of the selected set of assets. Asset-related information can be spatially analyzed to help identify trends or to determine impacts of proposed operations. Analysis results and trends can be displayed on a map to further assist in the decision making process.

Where are the Benefits?

- *Executive Decision Support* (by providing current reporting of asset condition levels and map-based views of the infrastructure assets).
- *Customer Service* (by enhancing the customer response level and the details of work orders in a graphical context from the dispatcher to the field worker, such as a call center).
- *Mobile Work Orders* (by putting current maps with asset details and that day's work orders directly in the service truck on a mobile device such as a tablet, ruggedized laptop, or PDA).
- *Capital Projects* Budgeting and planning of capital projects can be improved by supporting the analysis of various alternatives prior to design. Construction management can benefit from having current maps and asset details for the existing and proposed infrastructure in the project's geographic area of interest.
- *Operations and Maintenance Supervisors* can more easily

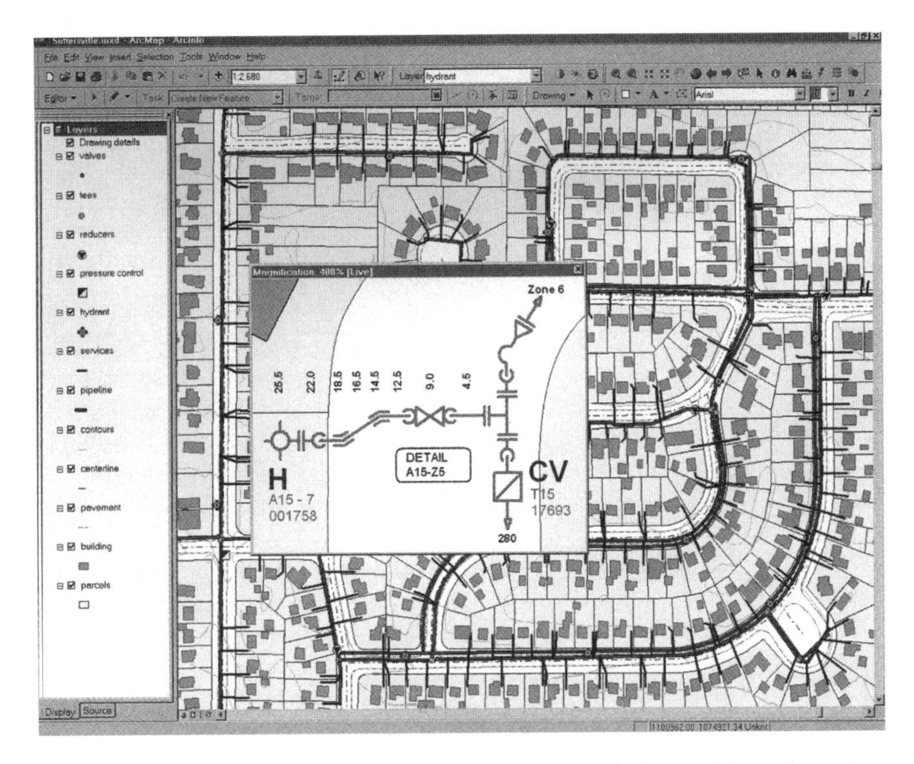

Figure 2 Using map-based views of the infrastructure to support decision making and reporting.

plan and manage field crews by knowing what equipment is needed and where. Maintenance workers can use daily or weekly work assignments that include routes and directions to the job, detailed reports of asset work history and condition levels, and equipment needed for the scheduled work.

- *Finance and Accounting* By tracking what work is completed where, budgets can be compared against actual costs of maintenance and repairs, cost overruns can be avoided through better planning and tracking of work and asset conditions, and continuous improvement of the infrastructure value can be achieved.

Why Bother?

- Cost savings to the organization can be realized by spatially enabling the asset management system.
- Often the investment in implementing a GIS has already been made.
- Extending the integrated system throughout the enterprise, including field services, offers many additional benefits to the organization.

Obstacles to Implementation

Unfortunately, there can be many reasons not to geospatially enable an AMS. Most of those obstacles are based on the total cost to implement, security concerns, lack of data standards (or the availability of metadata), complexity of operation (real or perceived), difficulty in coordination and cooperation between multiple groups, and resistance to change.

Many of these obstacles can be overcome by using good project management principles and by incorporating a change management approach that minimizes the risks of introducing new technology into an organization. The downside of not implementing such a solution must also be considered as part of the long-range strategic planning process for any organization that is asset-intensive in its operations.

> *"In the old world that is passing, in the new world that is coming, national efficiency has been and will be a controlling factor in national safety and welfare."*
> —Gifford Pinchot

Conclusions

By integrating the spatial component using a geographic information system (GIS), the asset knowledge base inherent in an organization's asset management system (AMS) can be better managed, shared, and visualized. Maps with current information become readily available to everyone from field workers to upper management.

Most of the obstacles to implementing such a solution are not difficult technical challenges, but rather require a good change management approach. The changes to the affected business processes that result from a geospatially-enabled AMS will likely lead to continuous process improvement, thereby reducing maintenance costs, improving customer service, enhancing operational efficiency, and increasing the return to the bottom line.

About the Author

Damon D. Judd is President of Ala Carto Consulting, a private consulting practice in Louisville, Colorado offering GIS and spatial data management services to utilities, energy, local government, and environmental organizations.

The Future of Imagery and GIS

By Adena Schutzberg

The highlight of the 2003 New England GITA (NEGITA) fall meeting "The Use of Imagery with GIS" was a panel discussion, moderated by Gerry Reymore of Early Endeavors, titled, "The Future of Imagery and GIS." The panelists included Gerry Kinn of Applanix, which is owned by Trimble, Ray Corson of James W. Sewell Company, and Gerald Arp who was with Space Imaging at the time and is now with Booz Allen Hamilton. Kinn provided an introduction by highlighting three trends in airborne imagery.

First, he noted the emergence of "designer" sensors. These are sensors geared to capture very specific types of data, in contrast to the more broadly used imaging and thermal sensors now used on planes and satellites. As an example, he cited a sensor under development at Rochester Institute of Technology that will detect very small fires. Such sensors will become more and more popular in the next five to ten years, Kinn argued.

A second trend Kinn termed "direct geopositioning," essentially the ability to create engineering level accuracy without ground control. Trimble is currently developing such technology. One group that might find such a technology disconcerting is surveyors. But Kinn was quick to point out that surveyors will always have the upper hand since they have the legal authority to interpret and use such data, just as today they interpret data from other sources, such as total stations and GPS receivers.

The third trend is the pent-up demand for GIS data. The graph illustrates a fast growing GIS marketplace. Growth, by Kinn's numbers, runs about 30% per year. On the other hand, remote sensing growth is just a bit better than flat. The gap between the two is pent up demand for GIS data that remote sensing is not filling. What other technologies are filling that gap? Other data capture technologies, particularly GPS, says Kinn. The remote sensing community is not delivering, due in part to the lack of tools for automated extraction of features (buildings, impervious surfaces, etc.) from imagery, he argues.

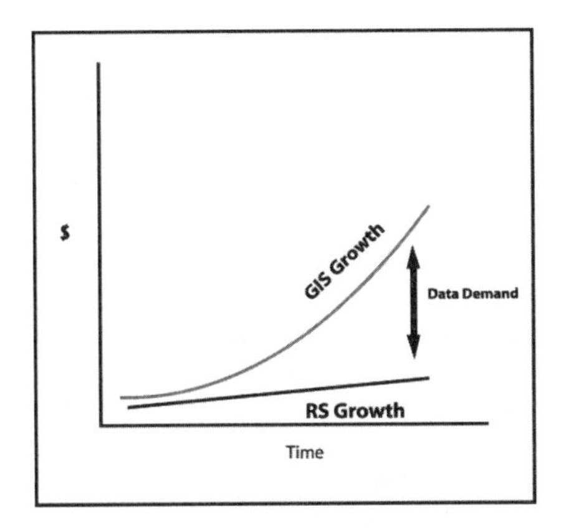

The gap between the growth of GIS and that of remote sensing can be considered the pent-up demand for GIS data that remote sensing is not yet filling.

That brought us to the questions for the panel:

1 What is the impact of federal mandates and activities on remote sensing?

ARP: The government has driven the commercial satellite remote sensing with the ClearView and more recent NextView contracts. These are essentially ways for the commercial sector to "fill in" for a dearth of spy satellites. The government is not severely limiting resolution as it's already granted licenses for ½ meter satellite images and made provisions for ¼ meter coverage. The government has also made it clear that it's trying to avoid sole source contracts (that is, providing a contract to a single vendor without competition) but prefers competition, which provides better quality, coverage, and price.

That said, despite the Open Skies policy, which allows freedom to capture data wherever one can fly, there are some government restrictions on what kind of data may be sold to whom (the denied persons list). For example, Space Imaging can't sell imagery with less than 2-meter resolution of Israel. And, the company can't sell imagery to Saddam Hussein and his peers. Interestingly, imagery vendors can provide data, for instance, of Cuba to Cuba.

KINN: Airborne imaging is like the "wild west" in comparison to satellite imagery. The latter receives government funding and has to obey lots of restrictions from government and nature. The airborne companies, in contrast, have limited funding: a million dollars is quite a lot of development money while satellite folks receive several million at a pop. We are "grubbing the low end." But, we do have some definite advantages: we can land, fix, or replace a sensor, develop a designer sensor, and we need not worry about Kepler's Laws of Motion or even high altitude clouds.

NGA [National Geospatial-Intelligence Agency, which was up until recently known as the National Imagery and Mapping Agency] is actively looking at airborne imaging to augment satellites, though that work is not as heavily funded. Still, there are some quick turn around solutions that should be available soon that will turn heads. For example, technology that allows the delivery of orthophotos as the plane lands may be available in the next year or year-and-a- half.

2 How do privacy issues play into imaging?

ARP: It's odd that privacy concerns are always raised when discussing satellite data, though aerial imagery has far greater resolutions and potential for capturing data that citizens might consider private. Satellite imagery is outside the threshold of personal privacy. Users, public and private, say they want higher and higher resolution, but in fact, few of the imagery purchasers want to process or receive that much data.

CORSON: We have not seen any problems with file sizes and resolutions. In fact, most of our clients are requiring higher pixel resolutions in an attempt to see more detail. Storage space has become inexpensive, making greater resolutions less of an issue.

KINN: It's not so much that we don't want to process or deliver the data; computers are fast and hard drives inexpensive. Instead the issue is the shear amount of raw data turned over to the agencies. As resolution increases, data size goes up as the square, so we are talking about huge data assets. The agencies can't go through it in a meaningful way.

As for privacy, Emerge did a lot of agricultural data capture. Those whose fields were included often asked if we'd be "sharing the data." They were concerned that they might get into trouble with the government since they were growing something they should not. They were particularly concerned that we were making a permanent record of their actions. Privacy goes out the window when you break the law.

3 What changes will we see in delivery—both methods and formats?

ARP: CDs, DVDs, and hard disks won't be going away too soon because Web-based delivery systems still have difficulties delivering very large data sets in reasonable periods of time. We've pretty much standardized on GeoTIFF for delivery. And, most of the government's special formats are losing favor.

KINN: Expect JPEG2000 to be a player soon.

4 With the world as it is, should we expect more government intervention in the coming months and years?

ARP: Actually, we should expect the opposite, as the government's already granting ½ and ¼ meter licenses. The government seems to have concluded that the benefits outweigh the liabilities. That may well be why the Iraqis put their materials underground.

CORSON: There are other ways to make data available but less useful for terrorists. It's possible, for example, to retain relative accuracy, but "play with" absolute accuracy.

5 Does the government have "better" imaging systems than the private sector?

ARP: I don't know and if I did I couldn't tell you. That said, TV and movies do exaggerate quite a bit. Thanks to Hollywood, we get calls from local police departments asking for an image of a particular street on a particular day, hoping we "caught" a robber running down the street. Besides, if someone wants higher resolution than we can provide, an airplane can usually do far better, assuming it can safely reach the geography of interest. One must presume that if the government has licensed commercial satellite companies to go to half-meter resolution, then their systems are still even better.

6 What should we expect regarding data packaging, delivery and licensing?

KINN: The aerial imagery marketplace is quite fragmented. It's tough to make "shrink wrapped boxed products" and sell them to many customers. What we do, essentially, is create custom products. And, to date, no one has figured out how to make 10 cents on each use of an image. We need to do what Kodak did with those little yellow boxes of film: make them ubiquitous and make a profit on each one.

As the entire imagery market matures, we are more likely to see a few bigger firms providing consumer level products/services. So, how will a future imagery company make money? Say sometime in the future you go to Wal-Mart to buy grass seed and fertilizer. New regulations limit the amount of fertilizer you can use on your property to cut water pollution. The salesperson uses a Web service to check out the size of your lawn via an image, then provides you with the right amount. You pay no more, but Wal-Mart sends a few pennies to the imagery company.

As for licensing, aerial imagery has a shelf life similar to bread. It's very tasty when fresh, but less and less interesting as it ages. Licensing will change to take that into consideration, but licensing will not go away.

ARP: Our licensing in the past was quite challenging. We've changed it accordingly so that our licensing is much simpler for "new" data. By the way, our images are already at Wal-Mart, in some of the video games. The new licensing takes advantage of the fact that in a few years the game maker will come back to us for up-to-date imagery for the cities covered in the games. In general our licensing is less and less restrictive, especially with the government customer.

CORSON: We actually have a historical film archive of many of the areas we've flown over the years dating back to the 1950s. A large percentage of our clients are interested in historical photography and mapping. The big challenge for us is being able to access this information digitally. We are scanning the archive film and creating metadata for it so we can find the images of interest in a massive project.

7 How will "designer sensors" play out?

KINN: Digital imagery is not being driven by remote sensing, but rather by other uses, such as taking pictures of Aunt Martha. What's happening in our corner is taking the off-the-shelf results of advances in other areas, and bolting them together to create "exotic" sensors. These might be used to sense specific targets like the ones currently in development at the Rochester Institute of Technology to sense tiny fires. Or, we might use multiple cameras to mimic the coverage and resolution of a single large "chip" that may be very expensive or not even exist. At some point sensors may be specialized enough to fly one mission and be retired. Of course, that's possible in aerial imaging, but not on satellites.

ARP: I don't agree with the designer view. In the commercial area the only change is in resolution, there is not a large push to add bands. To that end it is interesting to note that the latest satellite out of India, the P-6, [also known as ResourceSat] launched in October 2003 includes a green, red, and infrared band, but no blue! And there's no hyperspectral as there is not enough demand for it; it's too specialized for a commercial satellite.

The commercial market, which for us means primarily the government, is pushing exactly this type of trend.

8 How will LiDAR play into imaging?

KINN: Unlike imagery of a city, 95% of a DEM doesn't move over time. So, LiDAR doesn't require the refresh rate needed for say municipal applications. That said, "exotic" uses for LiDAR are on the horizon such as measuring crop height and canopy depth.

9 What can we expect when we need coverage of say a county, on demand?

ARP: Back a few years that was quite a challenging task. Over the years we've learned to "fly" the satellite and deliver those types of products. We are getting much better at that type of work.

10 What's the future of imagery prices?

KINN: Prices are certainly coming down. Other markets for "high tech" products suggest that specialized uses of technology give way in time to commercial uses and ultimately to consumer uses, with a corresponding drop in price. Consider what has happened with computers, for example. In GIS, we are dropping down to consumer products in services like MapQuest, but the business model for remote sensing is still unclear.

ARP: I agree that prices are coming down, but products will remain primarily custom. A few years ago we offered digital ortho quads. We expected to make a killing from people who wanted a picture of their house. Bottom line: it was a big yawn. We haven't found that niche. In point of fact, even our "standard products" are still custom.

KINN: The big issue is still that we offer data, raw data. We've not figured out how to offer the answer to the question, that is, information. That's why I think we keep coming back to the idea of automated extraction as a key tool to open the market. It can help change this raw data into information.

About the Author

Adena Schutzberg *is editor of* GIS Monitor, *a weekly e-mail newsletter from* GITC America, Inc. *She runs ABS Consulting Group, Inc. in Somerville, Massachusetts.*

Internet GIS: Power to the People!

A Web-based GIS provides a public-involvement tool for airport development

By Bernardita Calinao and Candace Brennan

A NEW AIRPORT RUNWAY CAN HAVE A MAJOR EFFECT ON A COMMUNITY. BUT HOW CAN CITIZENS BECOME MORE KNOWLEDGEABLE ABOUT SUCH ISSUES, AND HOW CAN THEY LET THEIR VOICES BE HEARD? IN ERIE, PA., C&S ENGINEERS USED A GIS-BASED WORLD WIDE WEB SITE TO ALLOW THE PUBLIC TO HELP CHOOSE WHICH RUNWAY EXTENSION ALTERNATIVES WORK BEST.

Erie's Internet GIS application is an analytical and public-involvement tool developed for an Environmental Assessment (EA) for the Proposed Runway 6-24 Extension at the Erie International Airport. The system is designed to help the public better understand the proposed project as well as its potential environmental and socio-economic effects. The process transcends the GIS from a tool for planners, managers and experts to its new function as a tool that enhances direct public participation for environmental decision making.

The basic elements of the Internet GIS reflect the contents of a standard EA, which include the following:

- Describe proposed alternatives.
- Present environmental feature maps within the project's area of influence.
- Delineate environmentally sensitive areas and present environmental consequences.
- Incorporate mitigation measures and action plans.
- The Internet GIS is consistent with Federal Aviation Administration (FAA) environmental guidelines, and it promotes the achievement of aviation-related environmental goals, including the following:
- Heighten objectivity in the EA process.
- Provide public information and input.
- Develop a more place-based approach to decision making.
- Ensure that the EA follows an iterative process.
- Comply with FAA's streamlining efforts.
- Develop new tools for interagency coordination.
- Create an effective platform for monitoring and environmental management.

The EA is developed through interagency coordination between the FAA and the Federal Highway Administration, which is currently overseeing the preparation of an EA for the relocation of Powell Ave., a road affected by the runway extension project.

Project Specifics

The Erie Municipal Airport Authority proposed a runway extension for Runway 6-24, the primary runway at Erie International Airport. The existing runway is 6,500 feet long and 150 feet wide. The proposed project would extend the runway 1,900 feet to the northeast. Safety issues are a primary focus of the proposed extension.

Via the Internet-based GIS, community residents can view the proposed project in the context of their environment. Users can identify tax parcels and relate them to the proposed runway extension and existing land use.

The EA process is evaluating the proposed environmental and socio-economic impacts of all the alternatives identified in the "Master Plan" so alternatives are evaluated using technical and cost considerations as well as environmental and socio-cultural considerations.

A Web site (*http://www.erieairportprojects.org*) was developed to facilitate the link among the Internet mapping system and other related environmental information. The Web site's principal feature is a section on "Environmental Maps," which carries all the features of an Internet GIS. Because the project is developed in conjunction with the Powell Ave. project of the Pennsylvania Department of Transportation, the Internet GIS may likewise be accessed at *http://www.airport-powellprojects.com* under "Project Details, Airport Information."

An Online Mapping System

The main objective for the Internet GIS is to provide public access to information used in developing the EA. Project engineers didn't want to spend much time processing and converting existing data. They also needed a product that could deliver a lightweight Internet application to the public as well as be accessed through a standard and readily accessible Internet client such as a Web browser.

Application Requirements

Features such as runway alternatives, noise contours, runway safety areas and runway protection zones were created in a GIS format. Using a GIS as a data management tool is effective, because it can overlay and query spatial data. Government agencies already had data needed by the project. Tax parcel information was used along with details of the runway alternatives to calculate the number of residential properties affected by the proposed changes. Other information was added to the system, including watersheds, hazardous waste sites, air-quality monitoring data, roads, land use, neighborhoods and socio-economic census data.

By using an aerial photo as a background, a more realistic view of the potential impact areas is provided.

Airport environmental planners considered several different ways to make such information available. If they released the data on a CD-ROM, they would quickly lose control of updates and additions. The project requires data to be centralized, and updates need to be instantly displayed. If the engineers created a kiosk for the public to use at a specific location, they wouldn't be able to service multiple users at one time. An Internet application can handle multiple users from different locations, and the application can be reused and customized to meet the needs of future projects.

From the public's perspective, there are many issues to consider. The application should be quick, easy and convenient to use for someone inexperienced with computers and mapping applications. The project's Internet GIS doesn't require prior GIS knowledge, and there's no need to download or buy extra software. There are different choices for the type of viewer. Some require Active X or Java plug-ins, while others use basic HTML and Java Script, which are lightweight and provide access to all levels of Web users.

The main Web site serves as a platform and provides additional information about the project and the Internet GIS.

Content is another consideration. The site can be used to re-create and investigate maps seen in public meetings. It can be viewed by the public or accessed by people working on the project if they need quick information. The citizens living in the neighborhoods surrounding the airport are interested in evaluating the effect of runway extensions relative to their homes.

In addition, the Internet GIS will give insight to questions such as:

- Where are the extensions proposed?
- Will I have to be relocated?
- Will this change the noise levels around my home?
- Will I be able to receive benefits from the Sound Insulation Program?

Internet GIS Configuration

The tool chosen to connect the public with the data was ESRI Inc.'s ArcIMS software, because most of the base maps and GIS layers were in ESRI's Shapefile format. The specific application directly accepts data that already have been collected and managed for other stages of the runway extension project.

There are three types of data presently included: 1) existing conditions, 2) potential impact areas and 3) cumulative effects. The existing conditions include all existing information such as roads, buildings, wetlands, floodplains, census block data, tax parcel information and airport pavement. Potential impact areas are created for each of the runway alternatives. The primary impact area is the predicted soil disturbance area and a buffer of approximately 500 feet around the runway safety area. The GIS also takes into account the runway protection zones defined for each alternative. Some cumulative areas will be affected by more than one source, such as a neighborhood block, for example, which will receive a significant increase in noise level and a reduction in air quality.

ArcIMS is a server-side software product that depends on a Web server and a Java servlet engine. Airport engineers already had a Microsoft Windows 2000 server with IIS 5.0 installed to serve company Web pages. There was enough room on the server to install ArcIMS, and the engineers networked another computer to serve the data. They installed The Apache Software Foundation's Jakarta Tomcat 3.2 product as a servlet engine as well as the ArcIMS application and spatial servers. The ArcIMS manager is installed on a separate computer used for development.

When a user views the Web site and sends a request for a map, the Web server sends the request through the Java servlet to the application server, which decides what to do with the request and sends it to the appropriate spatial server to handle the GIS computation. The spatial server then sends an output image of a new map to the user's Web server.

Web Site Design

Citizens need to see more than a map when they visit the Web site. Airport engineers decided to create a site that will introduce the project and EA process. This site includes sections such as "About the Project," "Questions and Answers," "Completed Environmental Reports," "Environmental Maps," "Other Erie International Airport Projects," "News," "Glossary," "Links" and "Public Comments."

The "Environmental Maps" section provides an area where prepared static maps can be viewed using Adobe Acrobat. In addition, the metadata for the layers used in the GIS application are available.

A simple HTML viewer was chosen for the application, because it didn't require users to install plug-ins. It's also the most lightweight in terms of processing required by a client's computer. All the spatial processing is done on the server side, and images are returned to the client. The viewer is interactive—it's able to accept requests and send responses back to clients. Tools available for users include zoom, pan, identify, overview map, view legend, layer control (on and off) and print.

All the layers redisplay images according to user preferences. The setup also allows users to adjust detail levels. The application provides more functions, but airport engineers decided to keep the format simple and only display the necessary options.

Project Concerns

Technical considerations are important when designing an Internet GIS. What type of audience is involved? What type of browsers will citizens have? Do they need an intuitive design or something more advanced? What type of equipment is available?

It's important to determine who is going to host the Web site. The airport project is served from two different locations. An Internet provider hosts the main site that includes the project descriptions, and the Internet GIS application is installed and hosted on a Web server at the C&S office.

The setup for the Internet GIS needs to be carefully planned before installation, and the existing infrastructure should be reviewed. It's important that network administrators work closely with GIS personnel to design an installation.

A significant issue that airport engineers faced when configuring their software was working around a firewall between the Web server and the rest of the C&S network. Due to some complications with strict firewall settings, the engineers decided to install all the ArcIMS components and data outside the firewall. Typically, this isn't the best installation, and the engineers are currently looking into alternative ways to configure the software.

Lessons Learned

The Internet GIS recently has been completed and released for public use in the EA process of the proposed runway extension. Although positive responses have been received to date, results from the participation effort still aren't available. A few implementation lessons, however, have become apparent.

For example, Internet GIS is a new tool for public involvement in airport development. As such, there's a need to stir more enthusiasm among airport regulators and environmental specialists. Airport engineers also experienced that, unlike a standard GIS, the political concerns are more pronounced in the use of Internet GIS, perhaps because information is made more available to the public. Therefore, strategic planning and quality control for any Internet GIS effort are extremely vital to effective implementation.

Map requests are sent from the client to the server, where the Web Server and ArcIMS applications handle the request. The request's result is an output image, which is transferred back to the client.

Internet GIS isn't intended to replace other formats of public participation. It's an approach that complements existing formats, because it allows citizens to access and interact with mapping and GIS data to enhance their knowledge about proposed land development projects in their community and increase their participation in the overall environmental decision-making process.

Internet GIS revolutionizes the way environmental assessment is conducted. As a result, it empowers citizens by providing them with a more dynamic and interactive tool for improved participation.

Calinao is senior environmental planner, C&S Engineers Inc.;
e-mail: bcalinao@cscos.com.

Brennan is a GIS specialist, C&S Engineers Inc.;
e-mail: cbrennan@cscos.com.

ORNL and the Geographic Information Systems Revolution

Explorers from competing teams race to find a mysterious lost city in the heart of Africa. The American team is continuously in touch with its Houston home base through satellite communications. In flight, team leader Karen Ross displays a map of Africa on her computer screen and notes the multicolored lines suggesting different routes from city to city and into the rain forest. Each pathway is accompanied by a precise estimate of travel time to the final destination. Zooming in on the target area, she switches to satellite images and interprets them in shades of blue, purple, and green. At each checkpoint, the team reports its progress and gets a revised estimate of arrival time.

Beset by difficulties, the explorers ask for a faster route, but the computer says the alternative is too dangerous. A simulation model with data representing geology, terrain, vegetation, weather, and many other geographic factors predicts local hazards, including the impending eruption of a nearby volcano. The Americans take the faster route anyway and beat the odds.

By Jerome E. Dobson and Richard C. Durfee

This fictional account of emerging geographic information system (GIS) technologies comes from Michael Crichton's 1980 novel *Congo*, which was made into a 1995 movie. The same technologies were highlighted in Clive Cussler's 1988 techno-thriller *Treasure*. In reality, GIS technology began more than a quarter of a century ago at key universities and government laboratories in the United States and Canada. Since 1969, Oak Ridge National Laboratory has been among the leading institutions in this diverse, now booming field. GIS has been evolving through new forms and applications ever since. Consider the following examples of GIS applications that rival and sometimes exceed Crichton's futuristic vision.

For the past three summers, ORNL geographers have monitored the potentially devastating effects of an Alaskan glacier with an annoying habit of rerouting whole river systems. We drive as far as the roads go or fly over roadless terrain with a color laptop computer that displays Ed Bright's interpretation of satellite images. A dot moves across the screen continuously showing our position on the image and thus on the ground calculated from Global Positioning System (GPS) signals from satellites. We've used the same system successfully in helicopters, boats, and even rental cars on the Oregon-Washington coast, the Gulf of Maine, and the North Slope of Alaska.

Since 1969, ORNL has been a leading institution in developing and using GIS technologies.

The roots of our remote sensing and GIS tradition started early at ORNL; more than 20 years ago, ORNL scientists studied some of the first satellite data from Landsat satellites (then called ERTS). By analyzing computer images of the Cumberland Mountains north of Oak Ridge, we were able to compute and display a three-dimensional perspective view of the

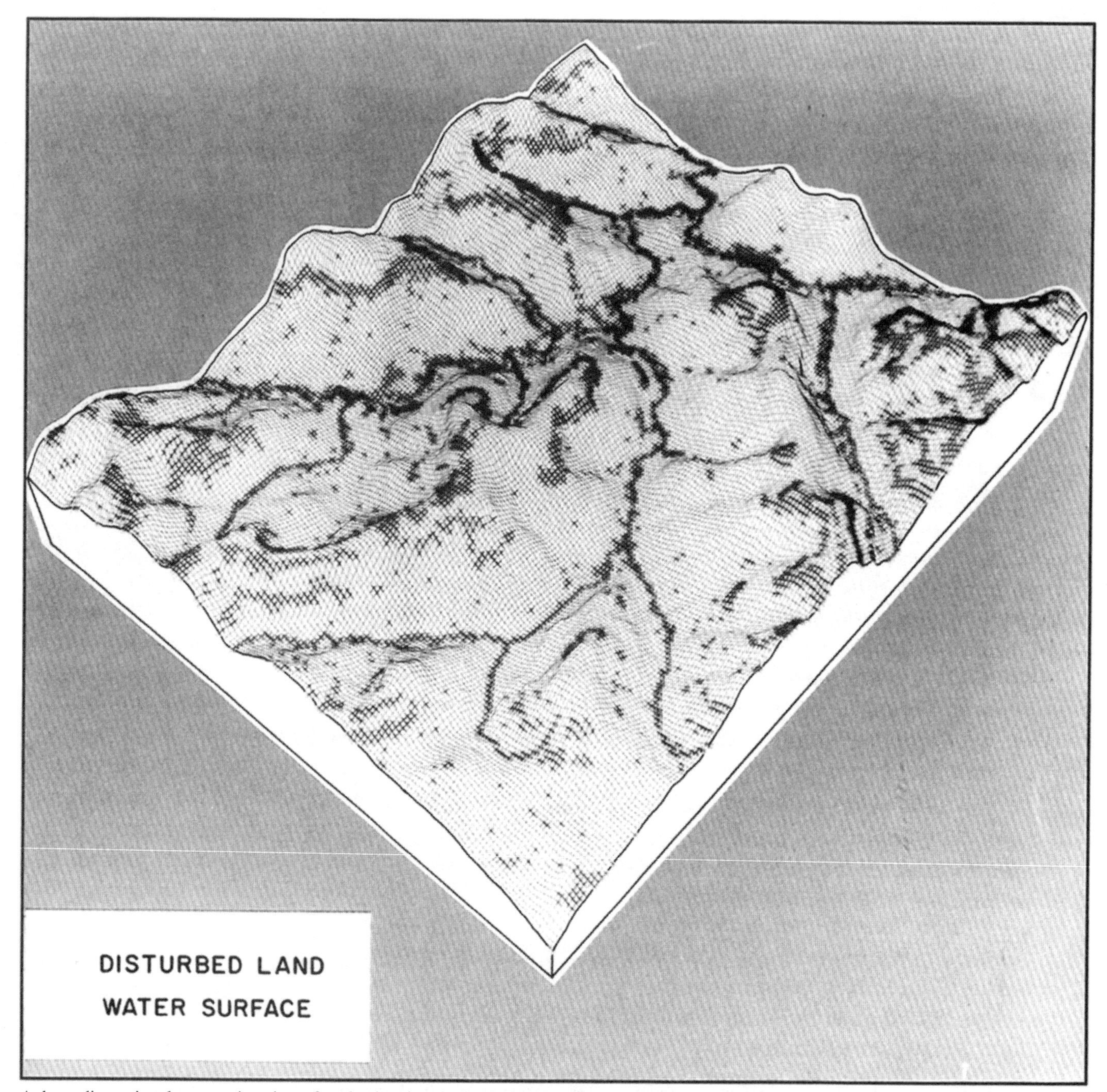

A three-dimensional perspective view of coal strip mines in the Cumberland Mountains north of ORNL. Streams are superimposed on the terrain. The spatial model computed the geographic relationship between disturbed land and nearby water surfaces, allowing assessment of environmental and visual impacts of strip mines.

coal strip mines in the area and superimpose the nearby streams on the terrain. After developing spatial models, we determined which streams were most likely to receive acid drainage from the strip mines. The visual impacts of strip mining on Oak Ridge residents were also predicted.

With or without satellite imagery, GIS is a powerful tool. In 1990, when the United States and other nations responded to Saddam Hussein's invasion of Kuwait, military leaders mounted the largest and most rapid deployment of military personnel and equipment ever attempted. The massive logistics were processed on the Airlift Deployment Analysis System (ADANS) developed at ORNL. ADANS, operating on networked computers, draws on a variety of logistic and spatial technologies to efficiently schedule the transport of U.S. military troops and equipment to trouble spots anywhere in the world. Since 1990, ADANS has been used to deploy military personnel and equipment not only to the Persian Gulf but also to Somalia, Rwanda, and Haiti.

In 1995, at ORNL's World Wide Web Showcase, Peter Pace showed a colorful high-resolution image of ORNL buildings and the roads, streams, and forested areas of the surrounding reservation. The view on his computer screen was constructed

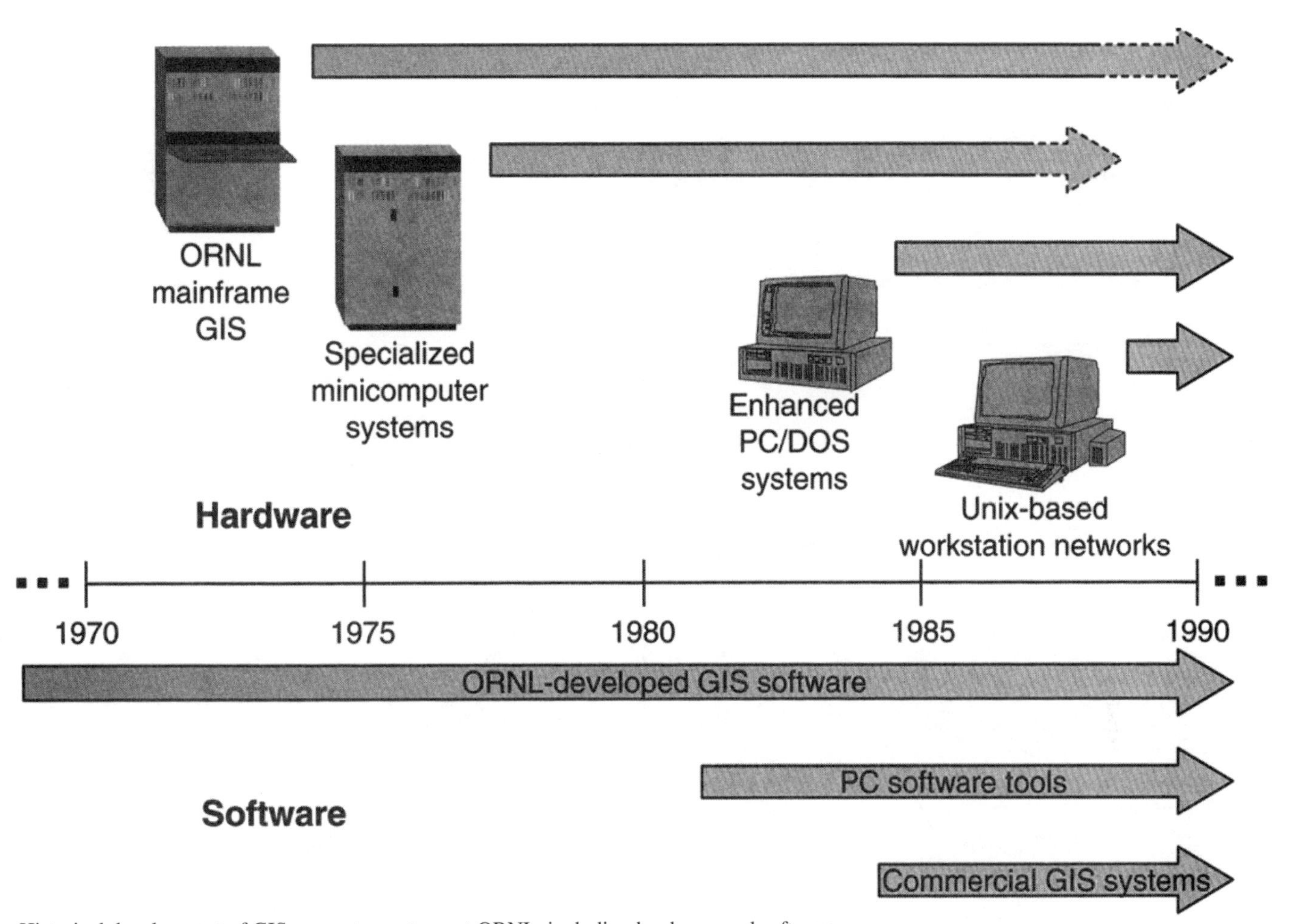

Historical development of GIS computer systems at ORNL, including hardware and software.

from a series of aerial photographs that had been scanned and converted to form a digital image. Various computer techniques were used to enhance and blend a series of images, eliminating unwanted elements and bringing out important details. Special photogrammetric techniques were used to remove distortions from the digitized photos. Each pixel (tiny rectangular element) on the screen represents 0.25 square meter (m^2) on the ground. Spatial registration of geographical features in the image is sufficiently accurate that a highly detailed map can be overlaid on the image. Pace zoomed in on a cooling tower and magnified it enough to see the blades of a fan. He printed out an image of the cooling tower alone. He and other ORNL researchers are preparing geographical data and imagery developed at Oak Ridge for distribution to selected users of the World Wide Web through Netscape, a navigational tool for accessing still and animated images as well as audio and text from the Internet.

Recent growth of GIS markets has been phenomenal. In 1994, GIS was listed under "Whole Systems" in the Whole Earth Catalog. Tens of thousands of people and organizations—universities, research centers, municipal planners, tax assessors, corporations, and resource managers—have come to depend on GIS for geographic data collection, analysis, and display. The commercial GIS industry, which started in the early 1980s, is now estimated to be worth $3.5 billion.

Today's rosy picture sharply contrasts with the situation in 1969 when GIS first began at ORNL. At that time only a few centers—principally Environment Canada, the U.S. Geological Survey, Harvard University, and ORNL—shared a common interest in solving the riddle of geographic analysis. Along with scientists from these centers and a few leading research universities, early members of ORNL's GIS and Computer Modeling (GCM) Group, led by Richard Durfee, contributed many of the developments that made the current boom possible.

Dobson served on a committee that composed most of the new national Spatial Data Transfer Standard.

These contributions include fundamental development of early geographic computational techniques that supported and accelerated the growth of a commercial industry; development and integration of key GIS data bases and methodologies; and use of geographic and spatial analysis to provide information to help policymakers make decisions on national issues, such as

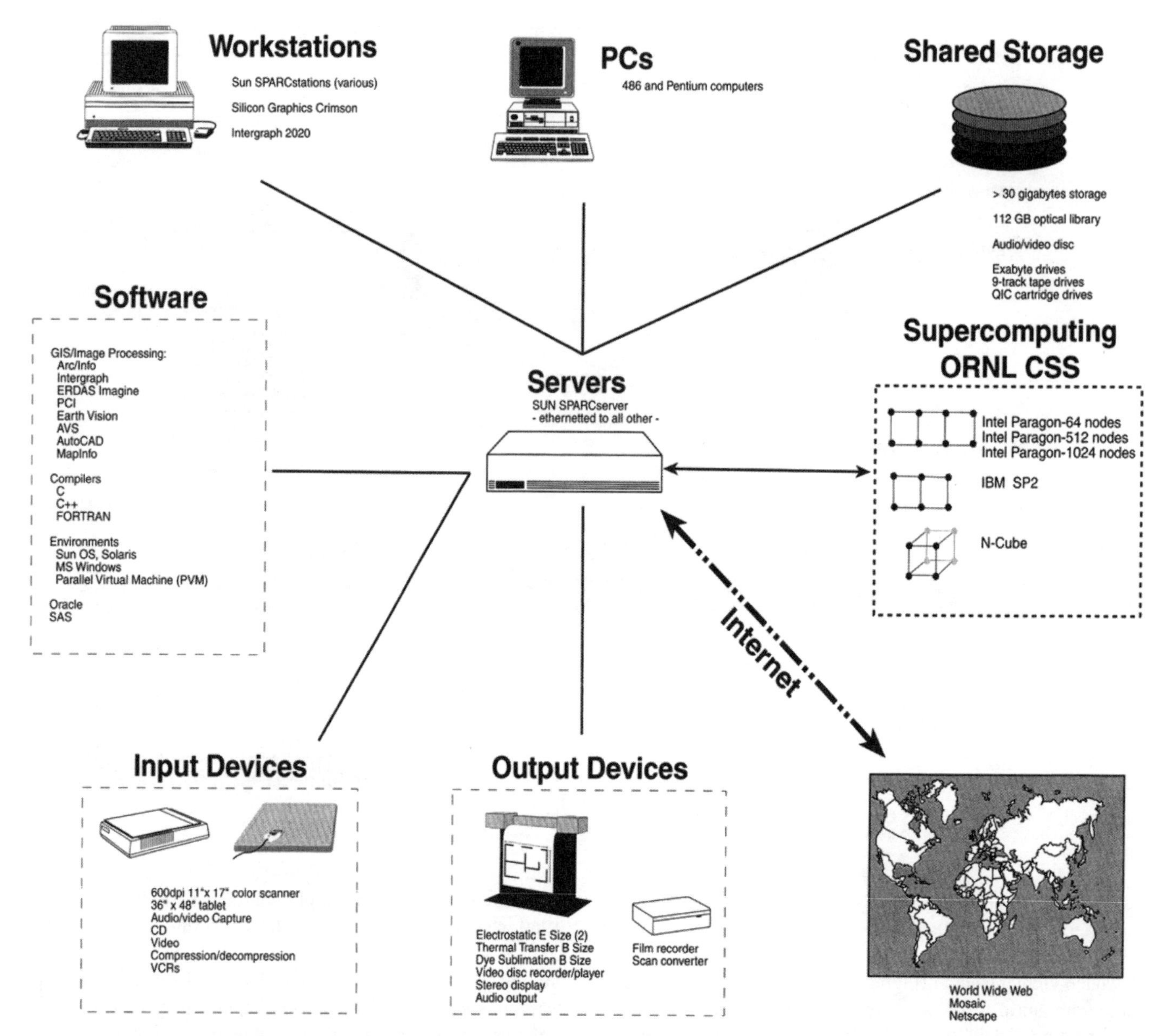

ORNL's GIS and Computer Modeling Group has a variety of computing resources used in support of a number of efforts ranging from natural resource assessments to environmental restoration.

development of energy sources and protection of water resources and fish populations, and to help government agencies assess natural resources and environmentally contaminated sites needing remediation.

Because of the increased use of GIS technology, a new national Spatial Data Transfer Standard (SDTS) has been established. Pioneering efforts by ORNL researchers Durfee, Bob Edwards, Phil Coleman, and Al Brooks helped build a foundation for the exchange of spatial data, and Jerry Dobson served on the Steering Committee of the National Committee for Digital Cartographic Data Standards, which composed most of SDTS. President Clinton's recently signed executive order requires all federal agencies to coordinate GIS data activities and make key data bases available to the public.

What Is GIS?

Many people think of GIS as a computer tool for making maps. Actually, it is a complex technology beginning with the digital representation of landscapes captured by cameras, digitizers, or scanners, in some cases transmitted by satellite, and, with the help of computer systems, stored, checked, manipulated, enhanced, analyzed, and displayed as data referenced to the earth. This spatial information includes earth coordinates and geometric and topological configurations to portray spatial relationships between features such as streams, roads, cities, and mountains. GIS is "a digital representation of the landscape of a place (site, region, planet), structured to support analysis." Under this broad definition, GIS conceivably may include pro-

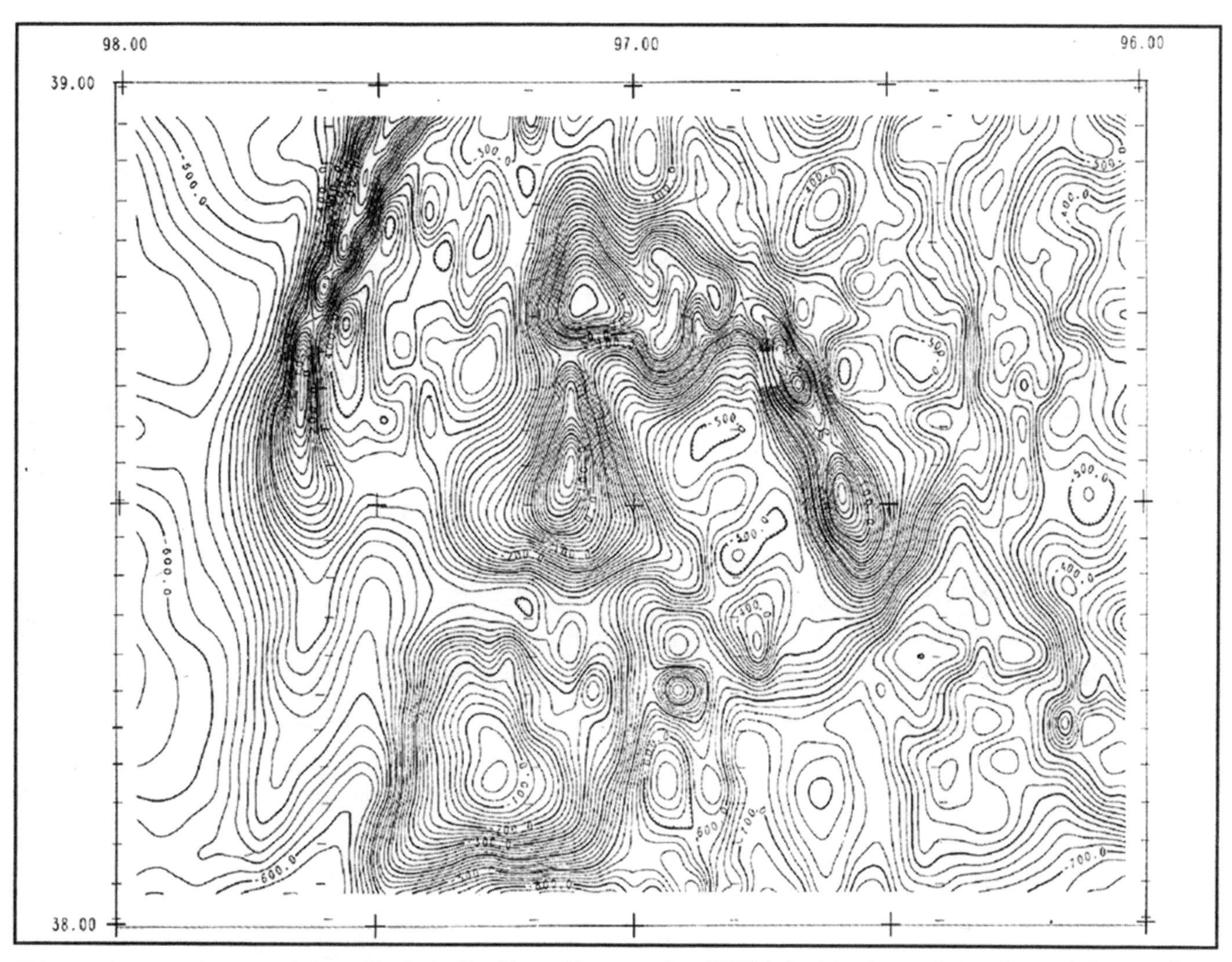

This map shows varying magnetic intensities in the Hutchinson, Kansas, region. ORNL helped develop a technique for converting one-dimensional flight line data into meaningful contour maps of regional magnetic data that could help identify anomalies potentially related to geologic deposits.

cess models and transport models as well as mapping and other spatial functions. The ability to integrate and analyze spatial data is what sets GIS apart from the multitude of graphics, computer-aided design and drafting, and mapping software systems.

Typical sources of geographic data for computer manipulation include digitized maps, field survey data, aerial photographs (including infrared photographs), and satellite imagery. Most image data are collected using remote sensing techniques. Aerial photographs are normally taken with special mapping cameras using photographic film. Most commercially available satellite imagery is collected using multispectral scanners, which record light intensities in different wavelengths in the spectrum—from infrared through visible light through ultraviolet light.

Spatial information can be represented in two distinctly different forms. Satellite images, for example, usually appear as raster data, a gridded matrix in which the position of each data point is indicated by its row and column numbers. Each position on a computer screen or map thus corresponds to the position on the ground measured by the satellite as it passes overhead. In contrast, cartographic features such as roads, boundaries, buildings, and contour lines usually are represented in vector form. In digitizing a lake, for example, the shoreline can be indicated as a series of points and line segments. In this case, each point is measured in Cartesian (X, Y) coordinates and each line segment is measured as a vector leading from one point to the next. The more points recorded, the more detailed the shoreline will be. Both forms, raster and vector, are essential to support environmental restoration projects on the ORNL reservation, for instance, and the software must be capable of rapid conversion from one form to the other.

GIS and remote sensing technology can detect changes in land features.

For such geographic information to be meaningful, it must be accompanied by "metadata" documenting the source, description, specifications, accuracy, time of acquisition, and quality of each data element. As GIS technologies and multitudes of geographic data bases have spread to the desktop in the past decade,

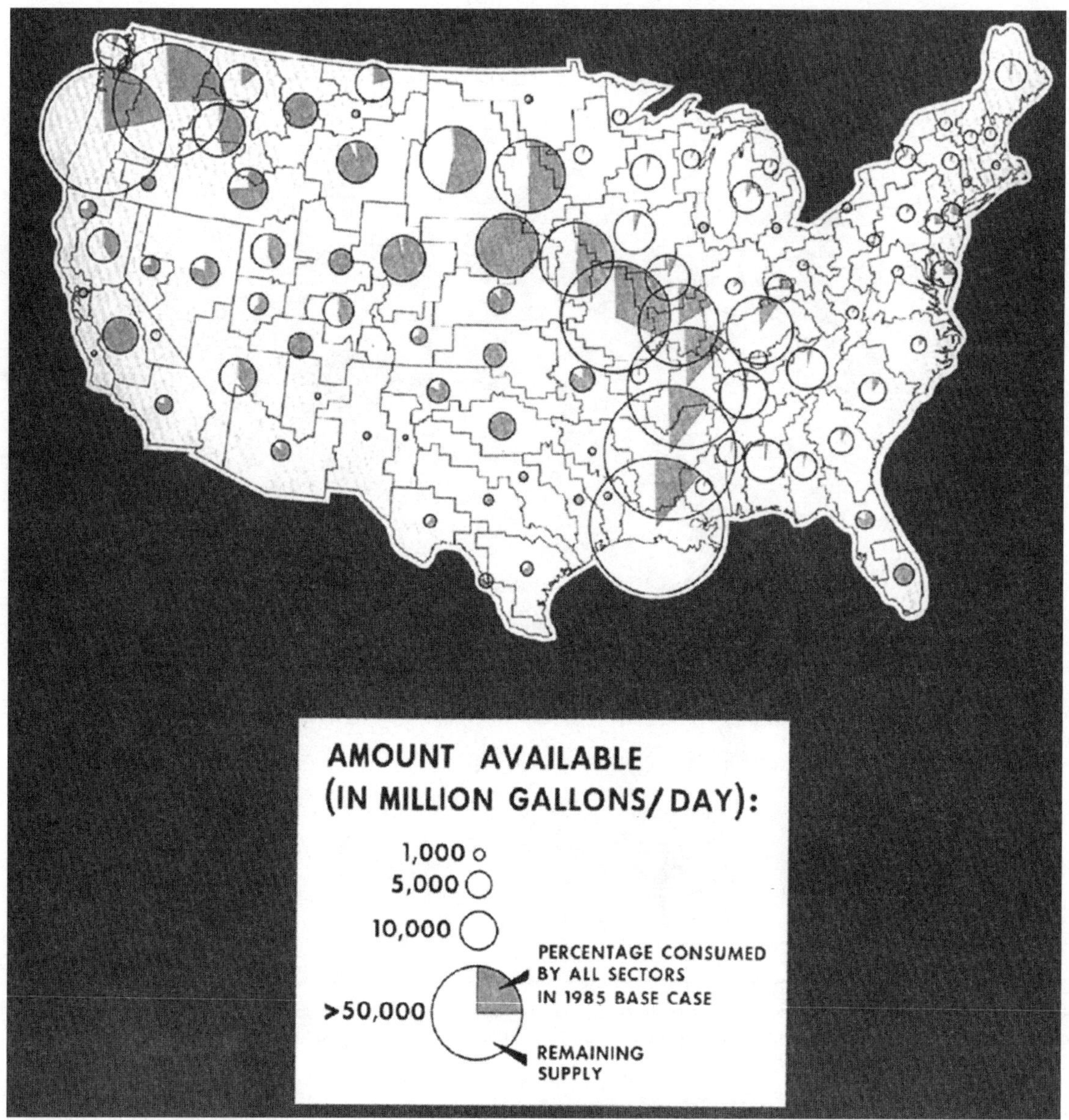

ORNL researchers used GIS and computer modeling techniques to predict the amount of water needed for national energy plans and compared it with the amount of water available in watershed basins across the United States.

metadata have become very important. Good metadata are essential in determining fitness of the geospatial data for each intended use—that is, determining which applications can be accomplished while ensuring the desired quality of results and decisions made from those data.

One of the most exciting applications of GIS combined with remote sensing technology is its ability to detect changes in features of large areas of land over many years by analyzing and comparing past and present landscape images. Each pixel can indicate a type of land cover, such as wetlands, forests, pastures, and developed areas. Such technology is now being used to monitor gains and losses in wetlands along the U.S. coast for assessing environmental impacts on U.S. fisheries. The technology has the potential for monitoring global change. For example, it is possible to detect increases in deforestation, which may alter the climate, or increases in desertification that may result from climate change.

In this article, we focus primarily on ORNL's role in the development and application of GIS to real-world problems over the past 25 years. Over this time, hundreds of projects and tasks involving GIS have been carried out by several organizations at ORNL involving a number of scientists, managers, and sponsors. It would be impossible to mention them all, but we do recognize and appreciate their significant contributions and collective vision for advancing GIS technologies over the years. In addition to the Computational Physics and Engineering Division, the examples of collaborating organizations within Martin Marietta Energy Systems have included the Energy Division, the Environmental Sciences Division, Chemical Technology Division, the Environmental Restoration Program, Biology Division, Data Systems Research and Development, and the Hazardous Waste Remedial Action Program. We highlight several of the larger efforts to illustrate the diversity of applications and tech-

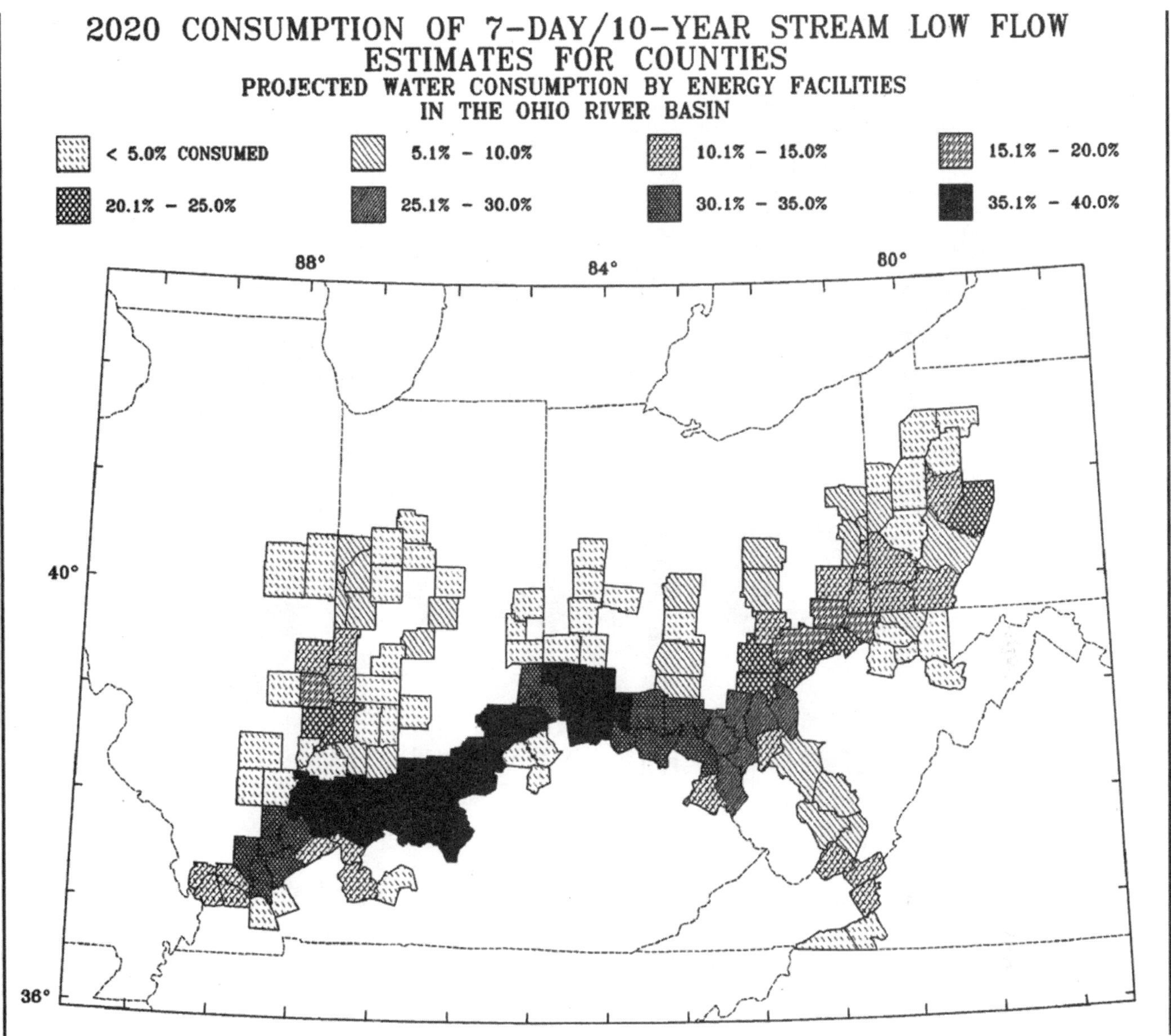

Results of one assessment of projected water consumption by energy facilities in the Ohio River Basin were shown to President Carter in a live presentation using a graphics station when he visited ORNL in 1978.

niques. We describe some of the early GIS developments and summarize some of the current systems capabilities. We offer examples in which GIS has proven useful in research and decision support.

History of GIS Development

Actually, the term GIS, though first introduced in 1964, was not extensively used until the late 1970s. The first comprehensive geographic data management system—called the Oak Ridge Regional Modeling Information System (ORRMIS)—was developed in 1974 at ORNL by Durfee. Its purpose was to integrate and support the data management needs of a series of regional analytic models depicting and forecasting land-use, environmental, socioeconomic, and sociopolitical activities in the East Tennessee region.

Many early ORNL developments in GIS that are commonplace today are remarkable primarily because of their dates. Examples from the 1970s and early 1980s include perspective and isometric drawings of cartographic surfaces, integration of remote sensing and statistical techniques with GIS, raster-vector transformation, viewshed calculation, polygon intersections, transportation routing models, and true three-dimensional (3-D) imaging.

ORNL has a long heritage of GIS research, development, and application to complex problems ranging from national issues to site-specific impacts. After presenting an overview of GIS technology development in ORNL's computing environment, we discuss three eras of GIS history at ORNL—regional modeling and fundamental development (1969–1976), integrated assessments (1977–1985), and issue-oriented research and analysis (1986–1995).

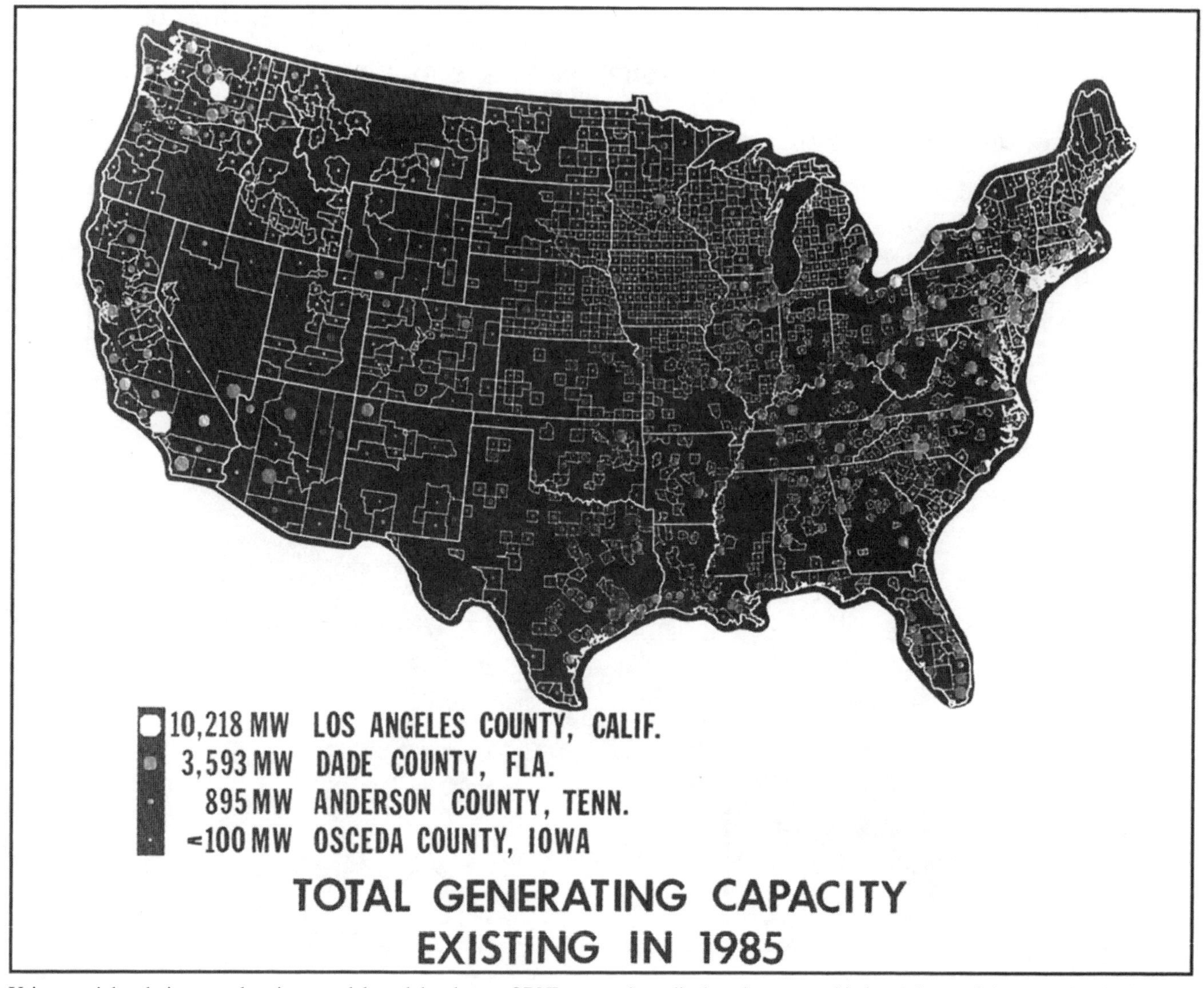

Using spatial techniques and various models and data bases, ORNL researchers displayed a geographic breakdown of the total electrical generating capacity in 1985 in the United States. The information aided utilities and energy agencies in the planning process for meeting projected demands for electricity.

Evolution of GIS Technology at ORNL

In the past 25 years, GIS software development and applications have migrated from mainframe computers to minicomputers to personal computers (PCs) to networked UNIX workstations. GIS software is now being modified for use on parallel processors and supercomputers, such as the IBM SP2 and Intel Paragon X/PS machines at ORNL.

Another technical development in the 1980s was integration of video information with digital data in the computer.

In the very early 1970s, a technological feat was the development of a computer-generated 3-D perspective movie by Tom Tucker of ORNL. The movie simulated terrain and population changes over a 40-year period in the Norris, Tennessee, area as Norris Dam began operation as a hydroelectric facility.

Over the years, one of the benefits of these spatial technologies has been their applicability to many different types of problems. One example was the development in the early 1980s of electron microscope tomography for 3-D reconstruction of DNA chromosomes as a collaborative effort led by Don Olins and his colleagues in ORNL's Biology Division in cooperation with the GCM group at ORNL. Adaptation and development of hardware and software for a commercial remote sensing system, I^2S, on GCM minicomputers played a major role in the analysis of electron micrographs and display of chromosome structures. When it was determined that more sophisticated true 3-D displays were needed, a special varifocal mirror display was built. Depth visualization was provided by a vibrating mylar mirror synchronized with a monitor mounted above the mirror whose image was reflected to the operator. Data at greater depths were displayed when the mirror was at a greater

This GIS image of part of the Adirondack Mountains shows the geographical relationship between forested areas where trees were blown down by strong winds in 1950 and acidified lakes (numbers on lakes show acidity levels, or pH). The forest blowdown has been linked to acidification of the lakes.

deflection, thus varying the focal length to correspond to the appropriate depth. This occurred at a rate of 60 times per second, so the observer saw a continuous 3-D image.

Another ORNL breakthrough in GIS technology in the mid-1970s was the development of vector-based algorithms and their eventual integration with raster-based grid cell systems. The GCM group used these techniques for all types of water-resource and energy-related studies in collaboration with the Energy Division. In the late 1970s, ORNL developed transportation data bases and capabilities for routing hazardous wastes across the United States. Through use of GIS technology to match proposed routes with population density, the health and safety risks of hazardous waste transport could be estimated.

Development of new GIS hardware and software. . . improved our ability to solve old problems.

Another technical development in the 1980s was integration of video information with digital data in the computer by Steve Margle and Ed Tinnel at ORNL. Raster digitization of video signals and the introduction of laser video discs opened up a whole new way of dealing with graphic and map data. Working in cooperation with the Data Systems Research and Development organization, the GCM group demonstrated the feasibility of using video from scanned map images recorded on laser video discs for simulations of war games as training exercises on a high-resolution workstation. In this technique implemented by Beverly Zygmunt, multiple video frames were located, computerized, and combined into large electronic maps that could be roamed and overlaid with other geographic and military information in real time.

Throughout the 1980s and into the 1990s, development of new GIS hardware and software technologies made new applications possible and improved our ability to solve old problems. It is interesting to note that some of our primary GIS applications in the first half of the 1990s have addressed a legacy of environmental problems, just as many initial applications in the early 1970s promoted GIS to help evaluate environmental impacts.

Some of the latest GIS research under way at ORNL involves developing software for use on parallel-processing supercomputers. Very recent work has shown that, by significant improvement of algorithms and by using parallel processors on ORNL supercomputers, the transformation and interpolation (estimation of values between data points) of large GIS data sets can be done 50 to 18,000 times faster than on smaller Sparc workstations. Because of the explosion in data collection from all types of earth sensor systems, workstations and supercomputers must be integrated to handle massive volumes of data.

Real-time airborne GPS techniques have been used in aerial surveys of the Oak Ridge Reservation.

We are also integrating portable GIS capabilities with GPS in which relative positions of objects on the earth can be pinpointed in real time by satellite sensors in communication with hand-held devices. As this technology becomes more commonplace, geospatial data will be collected at an ever increasing rate. Real-time airborne GPS techniques have already been used in aerial surveys of the Oak Ridge Reservation to collect high-resolution aerial photography with accurate positioning information. Computerized stereo techniques are being used with special goggles to help generate orthographic images (digital images corrected for camera, terrain, and other distortions) from stereo photography. Also, 3-D subsurface modeling and visualization are being done for hazardous waste studies.

To provide intelligent and efficient access to large amounts of geospatial data, work is under way to prepare and load this information on Internet and World Wide Web servers, which can be accessed by data browsing tools such as Mosaic and Netscape. These capabilities are important to the Oak Ridge user community and to the success of the National Spatial Data Infrastructure (NSDI) during the 1990s.

Regional Modeling and GIS Development (1969–1976)

In 1969, the U.S. Congress passed the National Environmental Policy Act (NEPA), the National Science Foundation (NSF) initiated the Research Applied to National Needs (RANN) program, Ian McHarg published *Design with Nature*, and ORNL delved headlong into regional modeling and GIS. Clearly, NEPA was a major impetus to the other three events.

Before NEPA, research and development, infrastructural development, and resource management decisions had been based almost exclusively on engineering and cost-benefit considerations. Suddenly, NEPA thrust all large enterprises, including the federal government itself, into a new legal and ethical milieu in which comprehensive, interdisciplinary analyses were absolutely essential. Alvin Weinberg, director of the Laboratory from 1955 to 1973, immediately recognized the need and sought to diversify the Laboratory's missions.

For GIS, the most important development in the early days at ORNL was the Oak Ridge Regional Modeling Information System and associated tools that supported spatial data input and display. The primary purpose was "to provide the data management capability for analysis models which forecast the spatial distribution and ecological effects of activities within a geographical region." The land-use modeling efforts became the principal impetus to remote sensing development as well as to the GIS expansions.

Initial GIS software techniques were based on hierarchical grid cell systems. It became apparent that additional capabilities were needed for accurate cartographic representation and analysis of vector-based map data. By the mid-1970s, development of sophisticated polygonal-based GIS systems at ORNL were well under way. Our development of efficient storage and computational techniques for integrating raster-based grid cell and vector-based systems opened the door to addressing larger and more complex problems with a national scope. Incorporation of new algorithms designed by Phil Coleman and Bob Edwards provided a capability for analyzing and displaying large national data bases.

Integrated Assessments (1977–1985)

In the mid-1970s, a shift in federal policy greatly reduced NSF funding for the DOE national laboratories. From then on, hardly another penny was received to support basic research, development, or operation of GIS systems at ORNL. The GCM group and the Energy Division shifted to applications-driven research, the funding for which allowed continued development and operations.

ORNL systems were used. . . to support conflict resolution in power plant siting.

We never had the luxury of focusing on a particular technology (remote sensing or computer cartography, for instance) to the exclusion of other technologies. We were then, and are still, comprehensive integrators with analytical purposes paramount in everything we do. In many respects, this approach has been advantageous because (1) the integrated GIS technologies were then applicable to a wide range of spatial problems, and (2) the applications-driven development minimized "ivory-tower" research looking for a problem to solve.

The first seeds of the new order were sown in 1975 when Richard Durfee and Bob Honea used ORRMIS tools for predictive modeling of coal strip mining and associated environmental problems. Results of this work were presented to Robert Seamans, head of the Energy Research and Development Administration (ERDA), predecessor to DOE. Soon afterward, we became heavily involved in siting analysis. In 1975 and 1976, ORNL systems were used, along with data from the Maryland Automated Geographic Information System, to support conflict resolution in power plant siting. By the late 1970s, these systems were heavily involved in decision support for federal energy policy and resource management. ORNL employed GIS extensively to evaluate the environmental impacts of various proposed National Energy Plans. Later, we predicted the amount of coal that could be produced from federally leased lands and evaluated the impacts on energy supply of designating certain lands as wilderness areas, thus protecting them from exploration for and extraction of oil, gas, and uranium.

During the mid-to-late 1970s, the Laboratory played a major role in the National Uranium Resource Evaluation (NURE) Program. ORNL's Computer Sciences Division (now the Computational Physics and Engineering Division), in cooperation with DOE's Grand Junction Office, was the national repository for all data collected and analyzed to assess the availability and location of potential uranium resources for future commercial nuclear power, research reactors, and other uses. ORNL staff were responsible for overall data management, GIS processing, spatial analysis, and mapping. Al Brooks was director of the Oak Ridge effort to support DOE in surveying the country for potential uranium resources and estimating possible reserves. Through a multitude of subcontractors, DOE conducted both aerial radiometric and geomagnetic surveys and hydrogeologic ground sampling on a quadrangle-by-quadrangle basis across the United States.

The aircraft had special sensors to detect radioactive isotopes of elements such as bismuth, thallium, and potassium as well as magnetic fields.

One example of highly specialized GIS work at ORNL was Ed Tinnel's development, in cooperation with Bill Hinze of Purdue University, of spatial filtering, interpolation, and contouring techniques to convert one-dimensional flight line data into meaningful maps of regional magnetic data. The purpose was to use these data to help study geologic features and identify magnetic anomalies that might indicate the presence of mineral deposits. These maps were also provided to the U.S. Geological Survey for publication. This was one of the earliest projects that required the handling of massive amounts of spatial, tabular, and textual information of many different types. During this time specialized GIS hardware systems were implemented to provide new ways of digitizing and displaying large amounts of geographic data.

Multiple energy assessments were early examples of policy analysis using GIS. A flurry of activity began each time President Jimmy Carter proposed a new National Energy Plan. Econometric models were run by the Energy Information Administration to project, as far as the year 2000, energy demand and fuel use by type in each major region of the country. These regional projections were passed to ORNL, where energy demand was disaggregated by Dave Vogt to Bureau of Economic Analysis Regions and supply was allocated to counties. Around 1980 Ed Hillsman and others of the Energy Division projected electrical generation from each existing plant and simulated construction or retirement of different plants by fuel type to determine if the president's goals would be met. Dobson and Alf Shepherd projected the amount of water needed for energy production and compared it with the amount of water available in each basin in the United States. ORNL's projections of electrical generation for different areas were passed to other national laboratories (Argonne, Brookhaven, Los Alamos, and the Solar Energy Research Institute), which used the information to evaluate effects on air quality, water quality, and labor supply. All results were reported to DOE, which conducted policy analysis of the feasibility of each proposed plan. Results of one GIS assessment of the projected water consumption by energy facilities in the Ohio River Basin were shown to President Carter in a live presentation using a graphics station when he visited ORNL in 1978.

In short, as early as the 1970s the nation's energy system and many pertinent physical and cultural features were simulated through GIS in linkage with econometric models, location-allocation models, environmental assessment models, and spatial data bases. The principal output was by county, but many of the data bases and computations covered details finer than the county level. For example, the data bases included population at the Enumeration District level, all power plants over 10 megawatts in generating capacity and all U.S. Geological Survey stream gauging station records. The models were as sophisticated as any in use at that time with or without GIS.

Another major multiyear effort involving ORNL researchers in the early 1980s was the development of a national abandoned mine lands inventory for the Office of Surface Mining (OSM) of the Department of the Interior. This effort, headed by Bob Honea, was based on federal legislation mandating that abandoned mine lands be reclaimed to protect human health, safety, welfare, and the environment, using funds collected as taxes on mining operations. A national inventory of abandoned mine lands was necessary to determine the affected areas in urgent need of reclamation and to establish priorities for reclamation of other sites. The effort was initially viewed as a technology-based project involving heavy use of remote sensing, GIS, record-based information systems, and statistical tools.

It was anticipated that analysis of Landsat satellite imagery would be a key ingredient for identifying detailed impacts from the disturbed, abandoned lands. However, an interesting turn of events made the project much more difficult than expected. When attempts were made to use results from satellite analyses to meet the mandates in the legislation, we found that the worst threats to human health and safety (e.g., open mine shafts, acid drainage, polluted water supplies) could not be determined from satellite data. Major environmental impacts could be addressed by analyzing satellite images, but health and safety impacts and reclamation cost estimates required field data collection and field assessment efforts. Thus, a major field collection effort, which included on-site interviews with affected populations, was carried out in conjunction with the state agencies of all the coal-mining states. Unique information handling techniques were devised to standardize and computerize textual, tabular, temporal, and spatial data from forms and maps that could then be linked with GIS for spatial aggregation, statistics, and mapping. Don Wilson was responsible for overseeing the computerization of all this information and development of a consolidated data base. These results could then support assessments at the state, regional, and national levels to aid OSM in allocating reclamation funds and overseeing mitigation of the severest problems.

Methodologies developed at ORNL for one application were readily adapted and applied to other problems. For example, our initial demographic work of the late 1970s was extended to compute detailed population distributions for any place or region in the United States.

The technique was used by Phil Coleman and Durfee to compute population distributions around all nuclear power plants in the United States. Our results, including the calculation of population exclusion zones, enabled the Nuclear Regulatory Commission to assess these exclusion areas—regions where additional nuclear power plants should not be built because too many people live or work there—to help make planning and licensing decisions.

Issue-Oriented Research and Analysis (1986–1995)

Starting in the mid-1980s, the emphasis shifted again, this time in a very positive direction, as GIS became an important tool in topical research on scientific issues of national interest, as illustrated in these four examples.

Lake Acidification and Acid Precipitation. Acid precipitation can cause water in lakes to acidify, potentially reducing fish populations. Lake acidification and other environmental issues that may be related to acid precipitation were major themes of GIS work at ORNL in the late 1980s. The Environmental Sciences Division (ESD) was involved prominently in the National Acid Precipitation Assessment Program (NAPAP), especially the National Surface Water Survey. Through extensive collaboration with U.S. Environmental Protection Agency (EPA) laboratories and numerous universities and private firms, Dick Olson, Carolyn Hunsaker, and other ESD personnel collected, managed, and analyzed massive geographic data bases for lakes and watersheds throughout the United States. The goal was to characterize contemporary chemistry, temporal variability, and key biological resources of lakes and streams in regions potentially sensitive to acid precipitation.

ORNL researchers concluded that forest blowdown facilitated the acidification of some lakes.

Simultaneously, the Energy Division approached the same problem from a different perspective. While NAPAP focused on impacts of acid precipitation, this project focused on watersheds and investigated possible causes of lake acidification.

In 1950, a huge storm with heavy rain and 105-mile-per-hour winds blew down numerous trees in 171,000 hectares of forest in the Adirondack Mountains of New York. In the 1980s it was observed that several lakes in the area were acidified, so one hypothesis was that the blowdown of the forest might be a cause. To determine if a relationship existed between the forest blowdown and lake acidification, Dobson and Dick Rush of ORNL and Bob Peplies of East Tennessee State University used an approach that combined GIS and digital remote sensing with the traditional field methods of geography. The methods of analysis consisted of direct observation, interpretation of satellite images and aerial photographs, and statistical comparison of two geographical distributions—one representing forest blowdown and another representing lake chemistry.

Associations in time and space between surface water acidity levels (pH) and landscape disturbance were found to be strong and consistent in the Adirondacks. Evidence of a temporal association was found at Big Moose Lake and Jerseyfield Lake in New York and at the Lygners Vider Plateau of Sweden. The ORNL researchers concluded that forest blowdown facilitated the acidification of some lakes by altering pathways for water transport. They suggested that waters previously acidified by acid deposition or other sources were not neutralized by contact with subsurface soils and bedrock, as is normally the case. Increased water flow through "pipes"—small tunnels formed as roots decayed—was proposed as the mechanism that may link biogeochemical impacts of forest blowdown to lake chemistry.

ORNL has led the technical effort to improve methods for analyses of changes in uplands and wetlands.

Both efforts illustrate an ORNL strength—the ability to assemble multidisciplinary teams and multiple organizations to attack complex problems. GIS, in itself, is an integrating technology because it draws together different sciences that have a common need for spatial data, visualization, and analysis capabilities. Such was the case in the acidification studies just described. Although primary responsibility for these two efforts rested separately in the Energy and Environmental Sciences divisions, the GCM group was heavily involved in both efforts. Thus, considerable interaction took place between the two projects. Since then, ESD, in cooperation with GCM and other groups, has continued to expand its GIS capabilities and resources. ESD scientists now have hands-on access to GIS systems and data bases to support a multitude of research efforts.

Coastal Change Analysis. For decades, the National Marine Fisheries Service (NMFS) of the National Oceanic and Atmospheric Administration (NOAA) has been concerned about declining fish populations in U.S. coastal waters. Suspecting that these declines might be caused by losses of habitat, such as saltmarshes and seagrasses, and increases in pollution resulting from expanding urban and rural development, as well as agriculture, NMFS initiated a research effort to solve the technical, institutional, and methodological problems of large-area change analysis—methods for determining the time, location, and degree of changes in large areas to better understand changes in ecosystems and ecological processes. ORNL has led the technical effort to improve methods for analyses of changes in uplands and wetlands, detected by satellite sensors, and to perform prototype satellite change analysis of the Chesapeake Bay. Integration of these remote sensing and GIS methodologies in a laboratory environment, in field investigations, in workshop settings, and for presentations and briefings in policy and management arenas shows how much this evolving technology is becoming ingrained in all phases of earth-sciences work.

The Coastal Change Analysis Program (C-CAP) is developing a nationally standardized data base of land cover and land-cover change in the coastal regions of the United States. As part of the Coastal Ocean Program (COP), C-CAP inventories coastal and submerged wetland habitats and adjacent uplands and monitors changes in these habitats over one to five years. This type of information and frequency of detection are required to improve scientific understanding of the linkages of coastal and submerged wetland habitats with adjacent uplands and with the distribution, abundance, and health of living marine resources. Satellite imagery (primarily Landsat Thematic Mapper), aerial photographs, and field data are interpreted, classified, analyzed, and integrated with other digital data in a GIS. The resulting land-cover change data bases are disseminated in digital form for use by anyone wishing to conduct geographic analysis in the completed regions.

Land cover change analysis has been completed for the Chesapeake Bay based on Landsat Thematic Mapper (TM) data. The resulting data base consists of land cover by class for 1984, land cover by class for 1988 and 1989, and a matrix of changes by class from 1984 to 1988–89. We found that, contrary to popular opinion, marshland in the Chesapeake Bay region increased slightly during the period. However, both forested wetlands and upland forests declined significantly, while land development expanded rapidly. At greater detail, we observed the formation of a new barrier island and recorded lateral movement of portions of its tip by almost a kilometer.

Although the Chesapeake Bay prototype focused on a single region, its purpose was to provide a technical and methodological foundation for change analysis throughout the entire U.S. coast. Four regional workshops (Southeast, Northeast, Great Lakes, and Pacific) addressed a full range of generic issues and identified the issues of special interest in each major coastal division of the United States. Ultimately, the protocol development effort involved more than 250 technical specialists, regional experts, and agency representatives.

During the summer of 1994, field work was conducted in the Gulf of Maine, along the Oregon and Washington coast, and in Alaska. The Alaskan study is especially interesting.

In 1986, the Hubbard Glacier moved, closing the narrow opening between the glacier and Russell Fiord's Gilbert Point on the coastline of Alaska. The ice dam later burst as the fiord's water rose, and the narrow opening was restored. The event was worrisome to salmon fishermen because the fiord's alternative outlet to the sea could destroy the unique stock of sockeye salmon that spawn in the Situk River. The glacier is poised to move again, and the new, more permanent ice dam that is expected could cause the fiord to empty through the Situk watershed, drastically altering its ecosystem.

Using satellite images of the Alaskan coastline from various years, we are identifying changes in the Alaskan coastline that will help predict the impacts on fisheries when the glacier closes the gap again. If the Situk River salmon are threatened, it may be necessary to transplant some of them to less vulnerable streams.

In studying satellite images, we have looked for changes in land cover from 1986 on and tried to quantify these changes on a regional basis. For example, we have looked at changes in the size and shape of woodlands, wetlands, grasslands, and bare ground over a period of years to characterize coastal changes. We are trying to model the direct relationship between land-cover changes and ecological processes.

To verify the accuracy of our interpretations of the satellite data, we visit the imaged sites. In 1993 and 1994, Ed Bright and Dobson went to Alaska to conduct field verification of a 1986 land-cover classification in the Yakutat Foreland and Russell Fiord. Now, when we do field work, we use a hand-held GPS device linked directly to a color laptop computer. Commercial software integrates the live GPS location coordinates with raster images representing land cover and with vector images representing other features such as roads. The device has more than doubled productivity in the field. We are currently designing a modeling approach that will link GIS, transport models, and process models to address the linkage between land-cover change and fisheries.

Environmental Restoration. To clean up a legacy of environmental contamination and to comply with environmental regulations, U.S. government facilities must locate, characterize, remove or treat, and properly dispose of hazardous waste. In the 1980s, ORNL researchers helped develop geographic workstations, spatial algorithms, 3-D subsurface modeling techniques, and data base systems for handling hazardous waste problems at Air Force installations. Later, this work provided a foundation for supporting environmental restoration activities at DOE facilities. Since the late 1980s, environmental restoration has become a major theme for GIS activities at ORNL. The integration of GIS with other technologies provides an important resource to support hazardous waste assessment and management, remediation, and policy formulation for environmental cleanup at DOE facilities. The locations of waste areas (i.e., surface operable units) across the DOE Oak Ridge Reservation (ORR) are represented by the bold polygons shown on the following map.

In studying satellite images, we have looked for changes in land cover. . . and tried to quantify these changes on a regional basis.

In conducting successful cleanup efforts and meeting regulatory requirements at these facilities, GIS can assist in many ways. Key aspects include investigation of the types and characteristics of contaminants; the location of possible pollutant sources; previous waste disposal techniques; the spatial extent of contamination; relationships among nearby waste sites; current and past environmental conditions, including surface, subsurface, and groundwater characteristics; possible pollutant

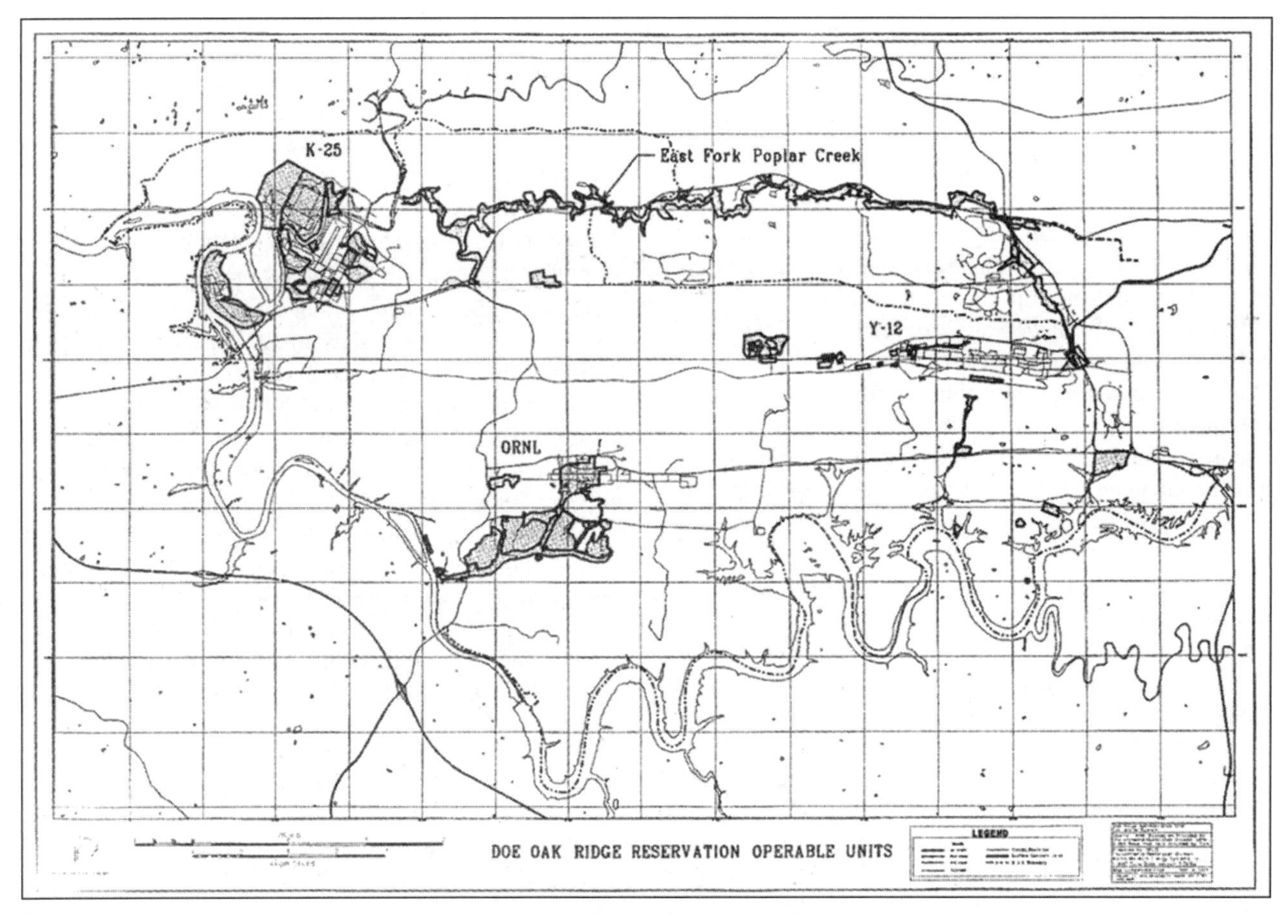

Waste areas designated as surface operable units on the DOE Oak Ridge Reservation.

transport mechanisms; efficient methods for analyzing and managing the information; effective cleanup strategies; and mechanisms for long-term monitoring to verify compliance.

GIS technologies support environmental monitoring and cleanup.

Three programs that involve significant GIS activities in support of environmental restoration (ER) in Oak Ridge include the Oak Ridge Environmental Information System (OREIS), the Remote Sensing and Special Surveys (RSSS) Program, and the GIS and Spatial Technologies (GISST) Program. The OREIS effort is designed to meet environmental data management, analysis, storage, and dissemination needs in compliance with federal and state regulatory agreements for all five DOE facilities operated by Lockheed Martin Energy Systems. The primary focus of this effort has been to develop a consolidated data base, an environmental information system, and data management procedures that will ensure the integrity and legal defensibility of environmental and geographic data throughout the facilities. The information system is composed of an integrated suite of GIS, relational data base management, and statistical tools under the control of a user-friendly interface. The OREIS effort, previously led by Larry Voorhees and Raymond McCord, is now being directed by David Herr.

OREIS's data have been combined with other site-specific information to study a variety of environmental problems. ORNL has modeled contaminant leakage from underground waste lines and used historical aerial photos to assess potential pollutant migration.

Results from such analyses are useful in locating potentially contaminated. . . areas.

The RSSS Program under Amy King supports ER site characterization, problem identification, and remediation efforts through the collection and analysis of data from aircraft and other remote sensors. One example has been helicopter radiometric surveys to determine gamma radiation levels across mapped areas of DOE facilities. GIS and remote sensing techniques also aid in the interpretation and visualization of airborne multispectral scanner data, thermal imagery, infrared and natural color photography, and electromagnetic and magnetic survey analyses. The following map shows examples of these types of processed information. Integrated results from such analyses are useful in locating potentially contaminated and affected areas, as well as possible underground structures that may be pertinent to hazardous waste burial and migration. Another example has been the delineation of waste trenches in burial ground areas that may be a source of waterborne contam-

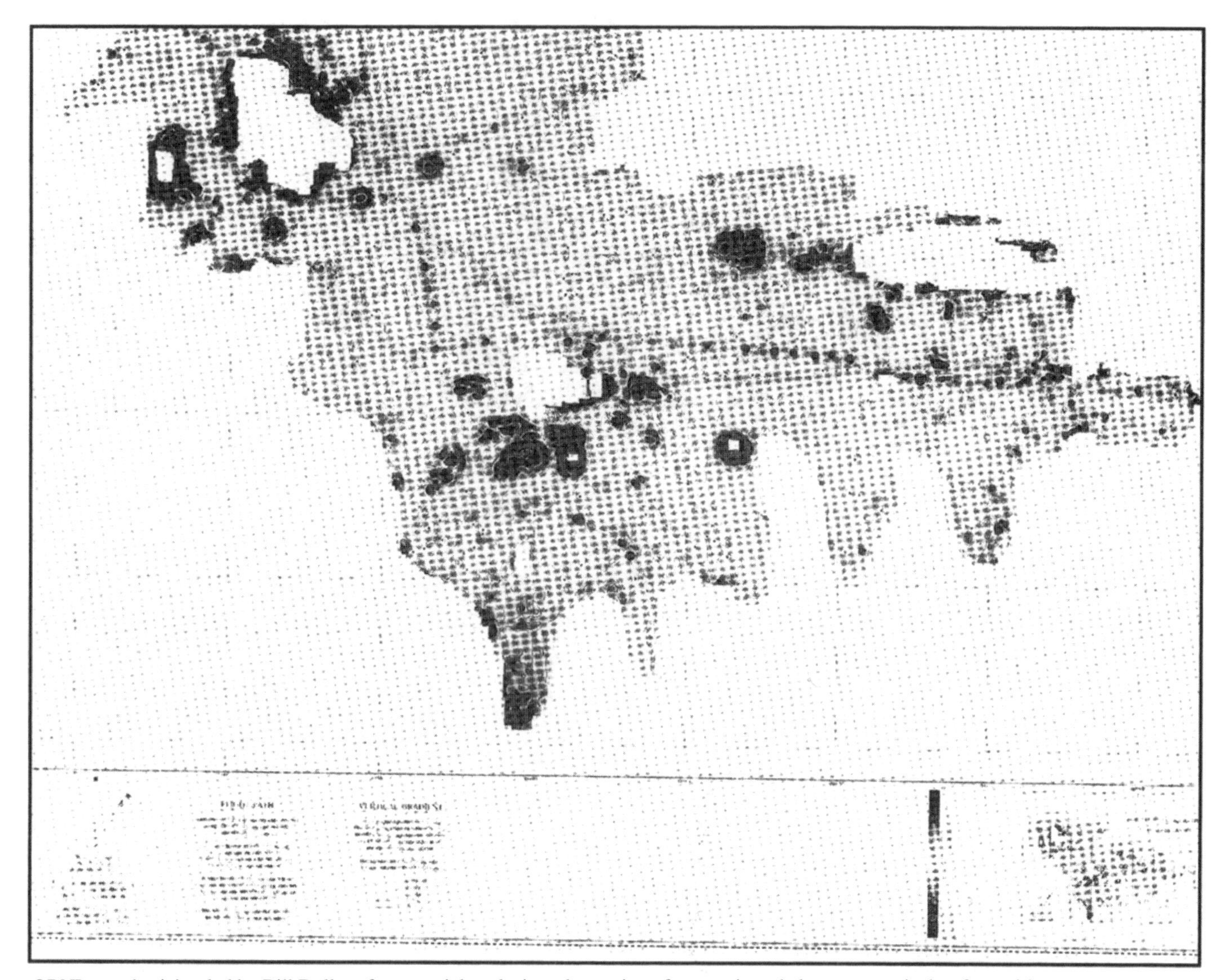

ORNL geophysicists led by Bill Doll perform spatial analysis and mapping of magnetic and electromagnetic data from airborne surveys to provide information that may aid in locating potentially contaminated areas on the Oak Ridge Reservation. Integration of this information with other types of remotely sensed data can help identify waste areas in burial grounds that may be a source of waterborne contaminants.

inants requiring remediation. The RSSS Program is also responsible for surveys of environmentally sensitive areas on the ORR.

GIS is used increasingly to plan, develop, and manage transportation infrastructures.

The GISST effort, under Durfee, promotes the development, maintenance, and application of GIS technology, data bases, and standards throughout the ER Program. The largest activity currently under way is the development of base map data, digital orthophotos, and elevation models for all Energy Systems facilities using advanced stereo photogrammetric techniques based on real-time airborne GPS. When completed, these terrain data will be the most comprehensive GIS and orthoimage coverages of any DOE reservation. This project, under the technical direction of Mark Tuttle, is being carried out in cooperation with the Tennessee Valley Authority. Desktop mapping systems are being integrated into the daily operations of many Oak Ridge staff devoted to monitoring and cleaning up the ORR. To support these activities, a repository of the resulting data from this project is being made available to users networked into a local file server, which will soon be accessible as a World Wide Web server. These GIS data provide a consistent, current, and accurate base map that can be integrated with all other types of environmental and pollutant data for analysis and reporting.

The fusion of all types of spatial data is an important tool for any environmental activity on the ORR. Through these and other ER programs, facility data and environmental data bases have been developed to improve understanding of relationships among pollutant sources, surface and subsurface pathways, and receptors of environmental contaminants. Three-dimensional modeling, data management, and contaminant analysis have

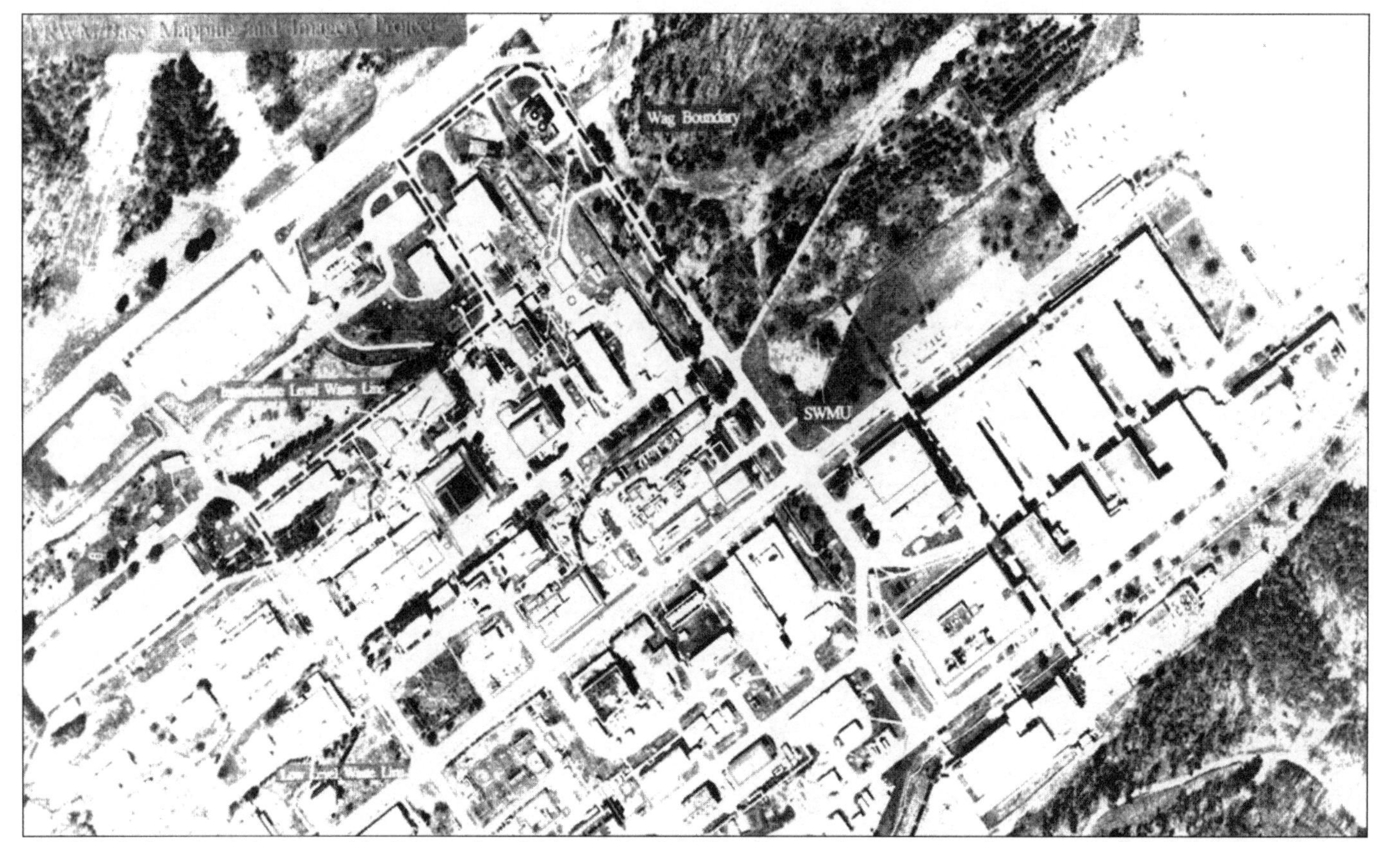

ORNL's fusion of orthoimagery with all types of spatial data on the Oak Ridge Reservation (note waste lines, operable unit boundaries, and solid waste management units in this figure) can help improve the identification and understanding of relationships among contaminant sources, support environmental monitoring activities, and assist in various types of facility management and land-use planning.

been enhanced through integration of computer tools and geospatial data. All these resources are becoming an integral part of the remediation planning and cleanup process, supported through communication networks linking scientists, engineers, and decision makers with analytical software and data bases.

Transportation Modeling and Analysis. Transportation systems and networks are crucial to the U. S. economy and way of life. GIS is used increasingly to plan, develop, and manage transportation infrastructures (e.g., highway, railway, waterway, and air transport networks) with the goal of improving efficiency in construction and operation.

Three main centers heavily involved in transportation modeling and geographic networks are the Energy Division (ED), the Chemical Technology Division (CTD), and the Computational Physics and Engineering Division (CPED). CTD has been primarily supporting DOE transportation needs in collaboration with CPED; ED has been supporting the Department of Transportation; and both ED and CPED have been supporting the Department of Defense. Collectively, the three groups have developed detailed representations of highway, railway, and waterway networks for the United States and military air transport networks for the entire world. ED, for example, is the developer and proprietor of the National Highway Planning System and the initial INTERLINE railway routing model. CTD has had a major responsibility for routing and assessing hazardous materials on the nation's highway and rail systems for many years (see figure next page). They have enhanced and adapted the INTERLINE and HIGHWAY routing models to assist in t his work. CPED has been a major developer of the Joint Flow and Analysis System for Transportation (JFAST), which is a multimodal transportation analysis model designed for the U.S. Transportation Command (USTRANSCOM) and the Joint Planning Community.

Operations Desert Shield and Desert Storm (1990–1991) involved the largest airlift of personnel and equipment from region to region ever accomplished. The U. S. Air Force's Military Airlift Command, now the Air Mobility Command (AMC), was responsible for this movement from the United States and Europe to the Persian Gulf region. Prior to that event, ORNL had worked with AMC to develop the Airlift Deployment Analysis System (ADANS), a series of scheduling algorithms and tools that enabled AMC to schedule missions to and from the Persian Gulf more rapidly and efficiently than ever before. ADANS is currently being used 24 hours per day by AMC to schedule peacetime, exercise, and contingency missions, as well as peacekeeping relief and humanitarian operations. Some of the key members of the ADANS team have included Glen Harrison, Mike Hilliard, Ron Kraemer, Cheng Liu, Steve Margle, and Irene Robbins.

The ADANS architecture is based on a relational data base management system, which operates on a network of powerful, UNIX-based workstations stretching across the United States with current installations at ORNL; Scott Air Force Base, Illinois; and

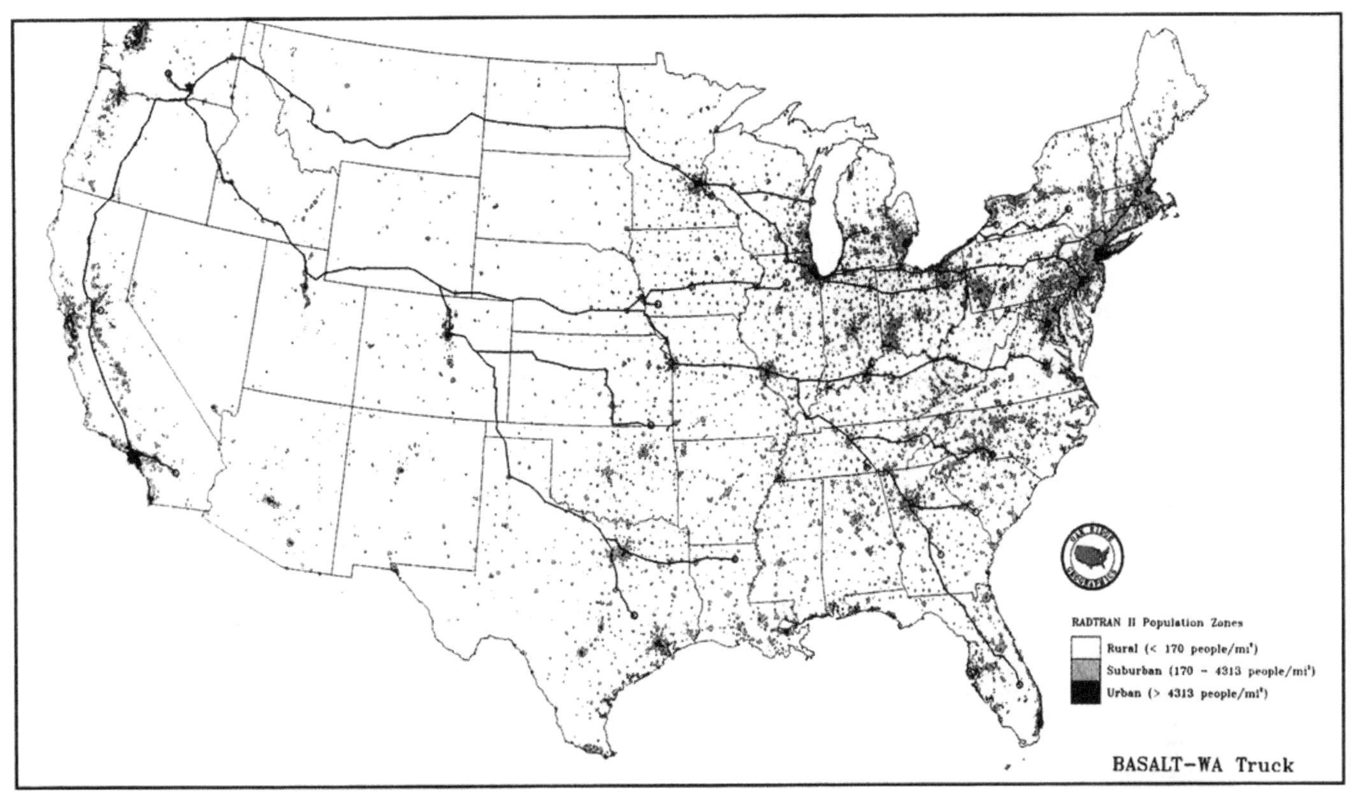

Results from Paul Johnson's and Dave Joy's work depict the computation display of possible truck routes for transporting radioactive waste material from groups of nuclear reactors to a candidate repository site. These routes are superimposed on zones of computed population densities across the United States.

Travis Air Force Base, California. PCs are used to perform some functions. The configuration includes a data base management system, a form generation tool, graphical display tools, a report generation system, communication software, a windowing system, and more than 500,000 lines of ADANS-unique code. All modules exchange data and run asynchronously. Thus, schedule planners can use the windowing system to keep track of and to modify multiple pieces of information. The three main components of the user interface are movement requirement and airlift resources data management, schedule analysis, and algorithm interaction.

All data and algorithms are geographically explicit. The user inputs data on a station-by-station basis with the textual network editor; the graphical network editor allows the user to establish a network and to enter or to edit information directly on a world map. With this system, it is easy to determine how cargo and passengers were moved, how many were moved as required, and to what aircraft they were assigned.

ORNL has advanced the use of GIS within our national infrastructure.

The JFAST effort, initiated by Bob Hunter, is designed to determine transportation requirements, perform course of action analysis, and project delivery profiles of troops and equipment by air, land, and sea. JFAST was used in Desert Shield to analyze the airlift and sealift transportation requirements for deploying U.S. forces to the Middle East and predict their arrival dates in-theater. These deployment estimates provided input for establishing concepts and timing for military operations. Under Brian Jones' direction during Desert Storm, JFAST was also used to track ships, provide delivery forecasts, and analyze what-if scenarios such as canal closings and maintenance delays. In addition to analyzing support for humanitarian efforts such as those in Rwanda and Somalia, USTRANSCOM and the Joint Planning Community use JFAST to determine the transportation feasibility of deployment plans.

JFAST incorporates a graphical user interface that makes significant use of geographic display of transportation data as well as other graphic displays to aid the planner in understanding the output from the flow models. To assist in preparing briefings, all JFAST screens and graphic displays can be captured and inserted directly into presentation software while JFAST is running. Data from JFAST can also be sent directly to other Windows™-compliant applications, such as spreadsheets and word processing packages.

ORNL's Role in the GIS Revolution

After a quarter of a century, how have GIS developments and applications at ORNL advanced science and served the national interest? ORNL has played an instrumental role in the GIS revolution by establishing and implementing a coherent vision that has been welcomed by scientific, policy, and management communities. ORNL has advanced the use of GIS within our national infrastructure. Today, commercial GIS products address many of the technical needs that required so much of our effort in the past, and the research frontiers have moved on to more complex methodological issues. However, no single commercial product today will handle all the current needs for GIS and related spatial technologies. One of our ongoing roles will be integration of multiple products with in-house technologies to best meet real-world needs that arise.

For the future, we envision linkages of GIS with environmental transport models and process models traditionally used by biologists, ecologists, and economists; implementation of GIS and digital remote-sensing techniques on supercomputers; 3-D GIS visualization and analysis; temporal analysis in a spatial context, and improved statistical analysis capabilities for geographic and other spatial data. The use of supercomputers will become even more important as new data collections—for example, the next generation of high-resolution satellite imagery—inundate the scientific community with terabytes of information. Justification for collecting and using these data will depend on the ability to extract meaningful information using supercomputer technology. We are currently addressing these and other technological issues, such as GIS animation, telecommunications, and real-time GPS and video linkage with GIS.

We hope to maintain a leadership position through continued advancement of hardware and graphics systems, GIS software, and data bases that will more effectively solve complex spatial problems. We think that knowledge-based expert systems will play a role in advancing future development and use of GIS technologies. We intend to assist the GIS community in improving standards and quality assurance procedures, and we look forward to assisting in enhancement of the National Spatial Data Infrastructure.

Ultimately, we view GIS as an integrating technology with the potential to improve all branches of science that involve location, place, or movement. Consider, for example, that most of the advances in medical imaging have been based on visual analysis. Imagine how much greater the potential would be if the images were enhanced by data structures, models, and analytical tools similar to those employed in analysis of the three-dimensional earth. We envision that certain technological thresholds will open the door to entire fields. For example, true 3-D analysis (more than visualization) and temporal GIS should provide new insights to geophysicists studying plate tectonics and the dynamic forces operating beneath the earth's surface. The single advancement of linking GIS with environmental transport models and process models will suddenly enable scientists and professionals in numerous other disciplines to incorporate spatial logic and geographical analysis alongside their traditional approaches. As these and other developments take place, a truly revolutionary new form of science should emerge.

BIOGRAPHICAL SKETCHES

Jerome E. Dobson is a senior research staff member in ORNL's Computational Physics and Engineering Division. He currently serves as chairman of the Interim Research Committee of the University Consortium for Geographic Information Science, scientific editor of the *International GIS Sourcebook*, and a contributing editor and member of the editorial advisory board of *GIS World*. He holds a Ph.D. degree in geography from the University of Tennessee. He joined the ORNL staff in 1975. He is a former chairman of the Geographic Information Systems Specialty Group, Association of American Geographers, and member of the Steering Committee of the National Committee for Digital Cartographic Data Standards. He previously served as leader of the Resource Analysis Group in ORNL's Energy Division, as visiting associate professor with the Department of Geography at Arizona State University, as a member of the editorial board of *The Professional Geographer*, and as a member of the Steering Committee of the Applied Geography Conferences. Dobson was co-founder and first chairman of the Energy Specialty Group of the Association of American Geographers. He proposed the paradigm of automated geography, and he was instrumental in originating The National Center for Geographic Information and Analysis and in establishing the Coastal Change Analysis Program (C-CAP) of the National Oceanic and Atmospheric Administration. Employing geographic information systems (GIS) and automated geographic methods, he has proposed new evidence and theory regarding the mechanisms responsible for lake acidification and regarding continental drift and plate tectonics.

Richard Durfee is head of the Geographic Information Systems and Computer Modeling Group (GCM) in ORNL's Computational Physics and Engineering Division. He is also program manager of the GIS and Spatial Technologies Program for Environmental Restoration supporting DOE's Lockheed Martin Energy Systems facilities. He is responsible for an advanced GIS Computing and Technology Center at ORNL with special facilities for analyzing and displaying all types of geospatial information. Previously, he was head of the Geographic Data Systems Group in the former Computer Applications Division and section head in the former Computing and Telecommunications Division. He joined ORNL in 1965 as a member of the Mathematics Division. He has an M.S. degree in physics from the University of Tennessee. An expert in GIS and remote sensing technologies, he served on the initial steering committee for the DOE Environmental Restoration GIS Information Exchange conferences and for the early DOE Interlaboratory Working Group on Data Exchange. He was also an early member of the Federal Interagency Coordinating Committee on Digital Cartography. He is coauthor of many publications and presentations, including a GIS-related presentation to President Carter during his visit to ORNL in 1978. For more than 25 years, he has researched and directed a wide range of GIS technologies supporting hundreds of applications for more than 15 different federal agencies.

From the *Oak Ridge National Laboratory Review,* September 1, 2002. *ORNL Review* is a U.S. Government publication.

Europe's First Space Weather Think Tank

A new program called COST Action 724 is poised to become the central coordinator of Europe's space weather research initiatives.

By Jean Lilensten, Toby Clark, and Anna Belehaki

Any space weather service must be able to give reliable predictions of the Sun's activity and its impact on the space environment and human activities. In Europe, as in North America and Japan, several private space weather initiatives have been launched in recent years, but daunting constraints—including the high cost of space-based monitoring and the vast complexity of space weather systems—make it imperative that large agencies coordinate the actions of researchers, data providers, users, and military organizations.

In Europe, this role is beginning to be filled by a new program sponsored by the European Cooperation in the field of Scientific and Technical Research (COST). A new program, informally known as COST Action 724, was recently approved to further develop the scientific underpinning of space weather applications within Europe while also exploring methods for providing space weather services to a variety of users. The organization, designed to coordinate space weather activities similarly to the NOAA Space Environment Center (SEC) in the United States, is an intergovernmental framework for coordinating nationally funded research among its 32 member states.

At the program's inaugural meeting on 24 November 2003, representatives from 17 countries elected Jean Lilensten as chairman and Anna Belehaki as vice chair for COST 724 activities over the next four years. Lilensten and Belehaki set several goals for the program's future, the most important of which is defining European standards for space weather data and services. They propose that COST should focus on initiatives for international space weather meetings, publications, and public outreach and education.

The chairs proposed as their first order of business to organize a meeting similar to the U.S. Space Weather Week with a European focus. Such an event would make it easier for many European industrial organizations to learn about space weather and help inform researchers that are currently involved in solar terrestrial programs that their research could fit in a wide-ranging space weather program. A European meeting on space weather would also encourage international scientific collaboration and promote public outreach about how their everyday lives may be affected by events in space.

Program Origins

The roots of the COST initiative can be traced to 1999, when the European Space Agency (ESA) initiated two parallel studies to examine the possibilities for a new space weather program in Europe. The two consortia, led by the Rutherford Appleton Laboratory in the United Kingdom and ALCATEL in France, involved a range of European industries and institutes. A team of independent experts called the Space Weather Working Team (SWWT) then assisted ESA in assessing the consortia's findings. Among the recommendations arising from these studies was a series of pilot projects for space weather services, including several that are currently under development.

One of the SWWT recommendations was creating a COST action to enable a space weather service to coordinate among various researchers, industries, and services. The COST technical committee for meteorology approved forming Action 724 in April 2002, and by the end of that year five countries—Finland, France, Germany, Greece, and Scotland had signed on. In 2003, Austria, Armenia, Belgium, Bulgaria, the Czech Republic, Denmark, Great Britain, Hungary, Israel, Italy, Poland,

Russia, the Slovak Republic, Spain, Sweden, Switzerland, and Ukraine joined the group, with ESA as an associated institution.

Other countries had until the end of May 2004 to sign on and claim part of the funding. COST actions, however, are designed to provide no more than 90,000 euros over four years. These funds, divided among all member countries, support only basic travel of committee members and other activities associated with coordinating research. Funding organizations in participating countries must provide monies for the research itself.

With little money at stake and no political ideology, the committee is like a space weather think tank. The motivation for a country to join a COST action is that these programs are known to set scientific standards for the European Union. Recommendations arising from a COST action, in addition, are likely to carry some weight in European policymaking organizations due to the international nature of the programs. An example of the program's influence is the creation of the European Centre for Medium-Range Weather Forecasts (ECMWF) that grew out of COST Action 70. Our vision is that COST Action 724 may eventually lead to the formation of a European equivalent—and partner—to the Space Environment Center in the United States.

News from the Inaugural Meeting

To achieve that ambitious goal, the organizing committee divided its members into four working groups, each targeting a particular segment of the space weather system. Attendees of the inaugural November meeting elected Mauro Messeroti of Italy and Werner Schmutz of Switzerland as heads of the monitoring and predicting solar activity group; Rami Vanio of Finland and Daniel Heynderickx of Belgium will lead the Earth's radiation environment section; Jurgen Watermann of Denmark and Stefaan Poedts of Belgium will pursue interactions between solar wind disturbances and the near-Earth environment; and Frank Jansen of Germany and Maurizio Candidi of Italy will focus on space weather observations and services.

It is important to note that two key space weather topics have been intentionally left out of the COST Action 724 master plan. The first, consequences of space weather disturbances on telecommunications, has already been intensely investigated under the guise of several earlier COST programs. Action 724 will also avoid the controversial issue of space weather's possible influence on terrestrial climate, in part because other groups, including the international Special Committee on Solar-Terrestrial Physics (SCOSTEP), are spearheading significant research efforts in that arena.

Next Steps

Enhanced communications among the working groups and with the public will also be a critical part of the committee's mission. To address that goal within the scientific community the working group plans to build an exhaustive list of the space weather programs currently available in Europe. Such an inventory can help to target points where data or services are missing or need augmentation. In addition, the organization will place a high priority on creating an easily accessible network among private and public space weather forecasting centers in Europe. The groups will also provide public information and advertising by coordinating a European space weather Web site and investigate the possibilities of initiating a European Space Weather Week.

Europe is gradually entering the arena of space weather. Because of its long tradition of public services, it is not surprising that the European organization of space weather research is driven primarily by governmental or multi-governmental agencies. The COST Action 724 initiative, thanks to its interministerial nature, has the ability to gather a large number of countries beyond the European Union and include industry along with collaborative and independent efforts. Its organizers fully expect that the new COST 724 initiative will play a central role in the European, and worldwide, landscape.

The newly elected leaders of COST Action 724 plan to present their findings on Europe's strengths and weaknesses in space weather at the European Geosciences Union (ECU) annual meeting in Nice, France, in April 2004.

Jean Lilensten *is a permanent researcher at the Grenoble Laboratory for Planetology in Grenoble, France.*

Toby Clark *was a project manager in geomagnetism with the British Geological Survey in Edinburgh, Scotland until 2002 and is now at the European Space Operations Centre in Darmstadt, Germany.*

Anna Belehaki *is a senior scientist and leader of the Ionospheric Group at the Institute for Space Applications and Remote Sensing at the National Observatory of Athens in Athens, Greece.*

Mapping the Nature of Diversity

A Landmark Project Reveals a Remarkable Correspondence Between Indigenous Land Use and the Survival of Natural Areas

Maps may be famously variable in accuracy, but generally speaking they are no more "objective" than are movies, novels, speeches, or paintings. Even if painstakingly accurate, they heavily reflect the interests of those who paid to have them made. Those interests may be political, commercial, or scientific. In the second half of the twentieth century, world maps emphasized the preoccupations of the Cold War, with a primary emphasis on international borders. The globes we had in our classrooms showed a world made up of nations. Until recently, most maps showed very little of what some of us now believe to be critical to the future of life: the boundaries of bioregions, watersheds, forests, ice caps, and biodiversity hotspots—and the principal ocean currents, wind currents, oceanic fisheries, and migratory flyways. In one of the offices at Worldwatch, there's a large map of North America showing nothing but the distribution of underground water. In *World Watch*, over the years, we've published maps of the global distribution of infectious diseases, war, slavery, refugee flows, and electric light as seen from space. The advancing technologies of Geographic Information Systems (GIS), combining the use of satellite imaging and digital data, have made these tasks easier by replacing laborious cartographic handwork with a capacity to superimpose maps of various elements showing how these elements may be related.

... [seeWorld Watch, March/April 2003, for a fold-out map that was] created under the direction of a nonprofit group called the Center for the Support of Native Lands, and produced in its final form by the National Geographic Society. [The map] was designed to exhibit two main categories of information: the distribution of cultural diversity in Central America and southern Mexico, and the distribution of forest and marine resources in that region. By superimposing these sets of information in detail, the map strongly confirms a hypothesis that has long been familiar to environmentalists and anthropologists alike: that there is a significant correlation of some kind between cultural diversity and biological diversity. That may seem obvious, as the homogenizing impacts of globalization are fueled and further exacerbated by a stripping of forests for cattle-ranching, plantations, and urban development. But in the past, the kind of data available to demonstrate this correlation on a regional or global basis has been fairly broad-brush. In 1992, for example, Worldwatch published a paper by Alan Durning, *Guardians of the Land: Indigenous Peoples and the Health of the Earth*, which included a diagram showing which nations had the highest cultural diversity (defined as those in which more than 200 languages are spoken) and which had the highest biological diversity (those with the highest numbers of unique species). Of the nine countries with the highest cultural diversity, six also ranked among those with the highest numbers of endemic species.

The history of Native Land's map of Central America and southern Mexico can be traced back even further, to the publication of the book *Regions of Refuge*, by the Mexican anthropologist Gonzalo Aguirre Beltrán, in 1967. Aguirre Beltrán noted that beginning with the Spanish conquest of Mesoamerica in the sixteenth century, indigenous peoples who had been decimated by warfare and by diseases against which they had no resistance sought refuge in "particularly hostile landscapes or areas of difficult access to human circulation." It was in those remote, often mountainous or jungle-covered areas that the refugees were able to rebuild their societies and preserve their cultures—and it is in those areas that they survive today.

In 1991, anthropologist Mac Chapin, who was then working for the Central America program of Cultural Survival, found himself perusing a map entitled "Indians of Central America 1980s," which had been compiled by the Louisiana State University Department of Geography and

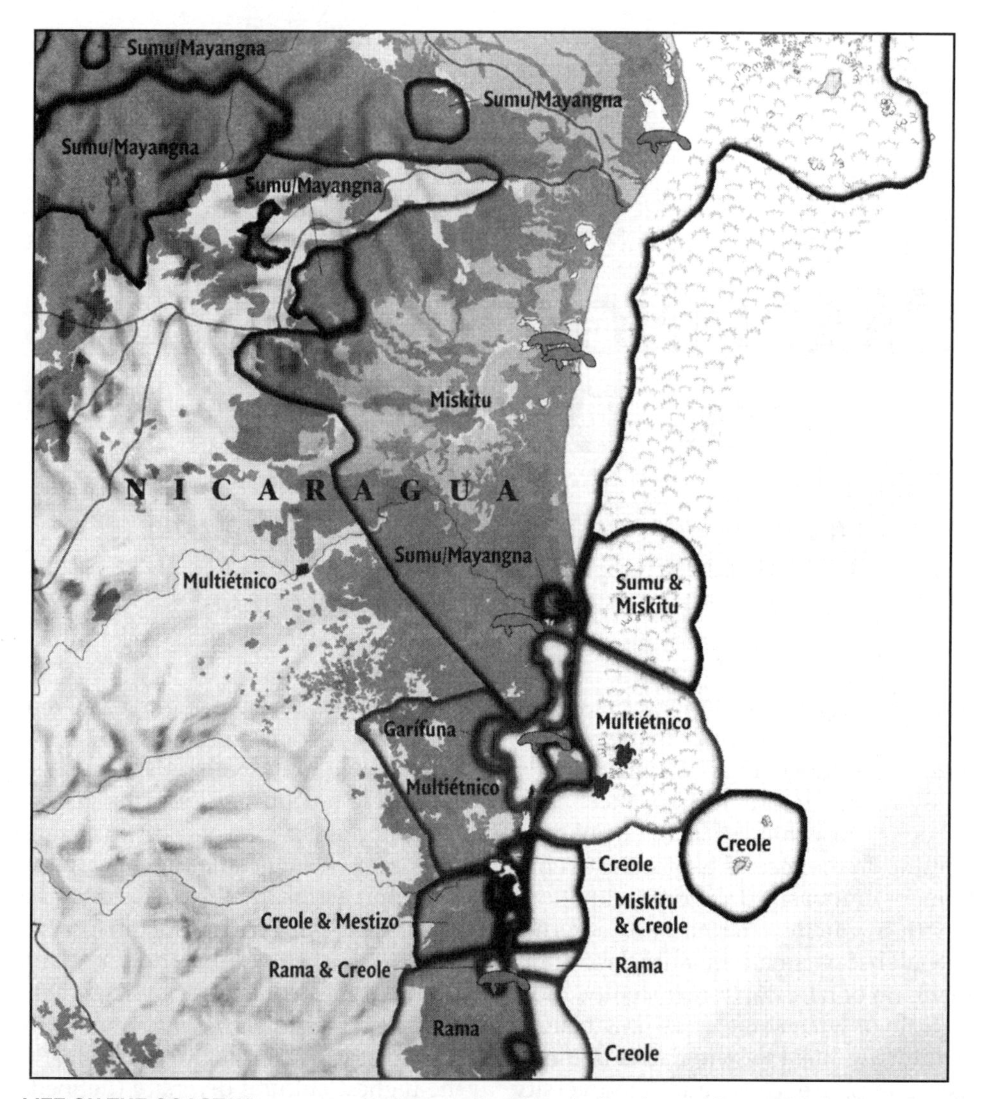

LIFE ON THE COAST When a map of indigenous territories (outlined areas) is superimposed on a map of forest cover (darker areas), the correspondence is close–as seen in this section of the Caribbean coast of Nicaragua. One reason for this correspondence is historic; the native groups retreated to densely forested areas centuries ago to avoid extermination by conquistadors. Another reason may be ecological, as the Indians' subsistence economy has proved less destructive to natural resources than has the "developed" economy. The ecological interdependence of forests and waters (estuaries delivering nutrients to the water, coastal land providing marine turtle nesting sites, etc.) is reflected in the mapping of indigenous territory, which is as much marine as terrestrial.

Anthropology two years earlier. He noticed that the indigenous population was located primarily in two areas—in highland Guatemala and strung out like a chain of beads along the Caribbean coast. "As I took this in," Chapin recalls, "I periodically looked at a 1986 National Geographic map of Central America hanging on the wall before me." The primary display was a standard "political" map, but in the corner was a small inset map showing, somewhat crudely, the region's vegetation. According to this map, most of Central America's natural forest cover was to be found hugging the Caribbean side of the isthmus—precisely where the lowlands indigenous peoples lived.

Chapin began thinking about the possibility of making a map showing the correspondence of indigenous settlement and forest cover, and shortly thereafter he was invited by Anthony de Souza, editor of the National Geographic journal *Research & Exploration,* to make one. The map—a precursor to the one which appears here [see "Life on the Coast"]—was published in that journal in 1992. It had small circulation but huge impact. Copies ended up on walls at the Inter-American Development Bank, World Bank, UN Food and Agricultural Organization, and the private residence of the president of Guatemala. One of the strongest impacts was on the indigenous peoples of the region. "The map helped to strengthen what soon became a widespread campaign for protecting and legalizing their territories," says Chapin.

That campaign also opened up a new venture for Native Lands—helping indigenous groups document their own land-use and marine-use patterns for purposes offending of incursions by developers, squatters, loggers, and the like. Traditionally, most of the native communities had regarded their territories as commons, and had never seen any need for such documents as plats and deeds; but the lack of such "proof" of ownership meant they were often unable to defend their territories from being occupied or exploited by outsiders. Chapin and his colleagues, who by 1994 had left Cultural Survival for Native Lands, embarked on a series of "participatory mapping" projects, in which indigenous groups made hand-drawn maps of their ancestral lands, and these maps were then combined with inputs from aerial photograph interpreters and cartographers to produce highly detailed, small-scale maps of the tribal territories. (One of the hand-drawn maps was published on the back cover of the January/February 1994 *World Watch.*) The progress of the participatory mapping program is summarized in a book coauthored by Chapin and his colleague Bill Threlkeld, *Indigenous Landscapes: A Study in Ethnocartography*, published in 2001.

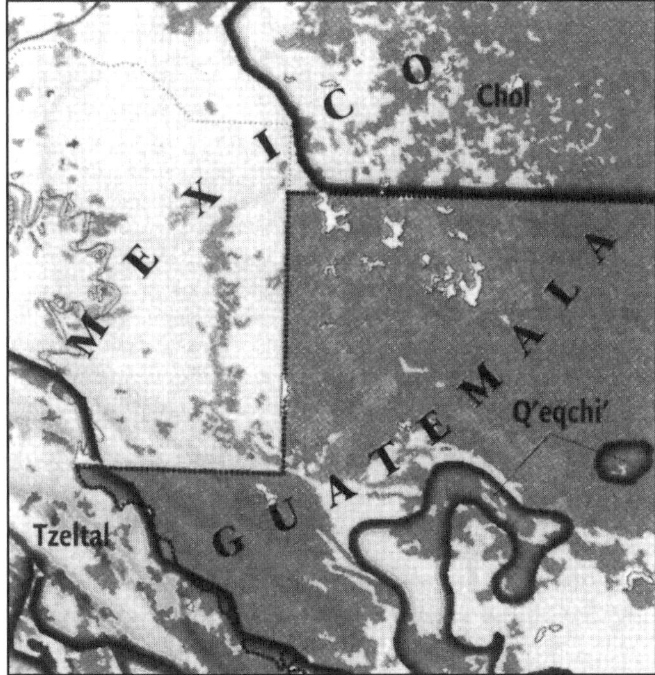

NATIONAL POLICIES, as well as cultural practices, can make a large difference in protection of natural assets. In this view, the border between Mexico and Guatemala demarcates a stark contrast between the land Mexico has allowed to be cleared for cattle grazing or timber, and the intact forest which remains across the border to the south and east.

In 2000, Native Lands decided to do an update of the original Central America map. "We knew that deforestation had advanced and that problems with Central America's Caribbean coastal environment—the bleaching of coral reefs and decline in fisheries—were increasing." Several heavily financed conservation projects had failed to halt the destruction. (The Mesoamerican Biological Corridor program of the World Bank had spent close to $100 million in the 1990s, and according to several assessments had little to show for it.)

It was also possible, by 2000, to make a far more accurate and informative map than the original. There had been major advances in satellite technology, and the success of the first version mobilized greater financial and professional resources. The original map had not included features of marine ecosystems, and had shown indigenous territories only on land. The new one would show marine-use areas, which are of critical importance to indigenous peoples along the Caribbean coast. The new map would also include the Maya region of southern Mexico, which the first map had not. The project would integrate the work of indigenous representatives, anthropologists, ecologists, and cartographers from every country from Mexico to Panama.

FROM THE MAP...

The first human footprints appeared along the isthmus that is today Central America and southern Mexico as early as 18,000 years ago. They were made by small bands of hunters and gatherers moving south through pristine landscapes abundant with plant and animal life. The newcomers prospered. They put down roots, and spread out, adapting themselves to the region's varied ecosystems. When the Europeans arrived at the end of 15th century, the native population was an estimated 7.680,000 people who spoke at least 62 languages and had cultural configurations ranging from tiny foraging tribes to the complex civilization of the Maya.

Contact proved disastrous for the indigenous peoples. As many as 90 percent of them died within the first 100 years, mainly from diseases against which they had no resistance. Most of those who survived retreated into the hinterlands to escape the unseen pestilence holding out in the remote fastness of the northern highlands and the humid forests of the Caribbean coastal slope. Indigenous peoples still have a strong presence in these areas today. Their numbers have steadily increased and now surpass pre-Hispanic levels with some 11 million people arrayed among more than 60 ethnic/linguistic groups. They are currently mounting campaigns to protect their ancestral homelands, natural resources and distinctive cultures.

Data-gathering took 15 months. The map was then designed by National Geographic Maps, the cartographic division of the National Geographic Society, and printed in an oversized (44 x 27-inch) format. The total cost, including data-gathering, design, and production, came to about $400,000. The map was printed with Spanish and English text and published as an insert to the February 2003 issue of the Latin American edition of National Geographic magazine, *National Geographic en Español*, with about 130,000 copies to be distributed in Central America and Mexico.

—*Ed Ayres*

From *World Watch*, Vol. 16, No. 2, March/April 2003, pp. 30-32.

MAPPING

Fortune Teller

For magazine readers of the 1940s, architect-turned-artist Richard Harrison mapped their world in a bold new way.

BY ANN DE FOREST

"THREE APPROACHES TO THE U.S.," READS THE HEADLINE. Underneath, printed in vivid color on a page of the old magazine, are three maps, three rounded edges of the globe, overlapping, like an illustration of a planet rising. In each, the land spreads out tan and yellow against the blue-green water. It almost seems that you're flying over this contoured terrain, about to swoop down for a landing. But what terrain is it? The headline says it's the United States, but these maps don't look like any America you know. Where is that familiar, iconic national outline? Where is Maine, waving in the upper right-hand corner? And where is California, bending forward like a ship's prow on the left?

The man who made these maps, Richard Edes Harrison, cartographer for *Fortune* for more than 20 years, specialized in such unorthodox presentations. He sought to, in his own words, "jolt... [readers]... with a new and refreshing viewpoint." So, on a Harrison map, north isn't always up, the earth isn't always flattened on the page and perspective—the point from which one views the planet—shifts, depending on what information he's aiming to convey. Says Joanne Perry, maps librarian and head of cartographic services at Pennsylvania State University. "He was trying to put out maps that showed truth, not convention."

In September 1940, the truth these particular maps revealed was far more ominous than refreshing. More than a year before the Japanese attack on Pearl Harbor, *Fortune* featured the images as part of an *Atlas for the U.S. Citizen:* 11 maps designed to show Americans they weren't as isolated from events in Europe and Pacific as they might like to believe. That swooping perspective, so breathtaking today, is actually the enemy's-eye view. Approached by air from Berlin, Tokyo or even Caracas, Venezuela, the United States suddenly looks vulnerable, with Detroit and Chicago tempting targets for the Germans, and Seattle and San Francisco poking perilously in Tokyo's flight path. And "if an enemy...," a caption warns, "should ever establish himself on the northern shore of South America or in the mazes of the West Indies," the Gulf Coast, hub of industry, becomes America's "soft belly."

Harrison, architect by training, artist by inclination, came to mapmaking accidentally (though he'd done some scientific illustrations for pay before studying architecture). Like many Americans in 1932, he needed work. A friend recommended him for an assignment to *Time* to make a map projecting the outcome of the looming presidential election. In 1936, he was kicked upstairs to another magazine in the Henry Luce empire, *Fortune,* where as part of its crack graphics team, he applied his visual skills to explicating the macro—charting the worldwide holdings of General Motors—and the micro—diagramming the inner workings of a gas mask.

World War II proved the ideal subject for his talents. This was a new kind of war, a truly global war, fought in the skies as well as on land and at sea. It was a war that demanded entirely new maps, new ways of seeing the world. Here, Harrison's lack of formal cartographic training was an advantage. Not bound by convention, he could make maps that opened readers' eyes to the realities of global geography in an aviation age.

"He explicitly saw himself as speaking to the general public, while he saw cartographers as failing in this regard," says Susan Schulten, a history professor at the University of Denver and author of *The Geographical Imagination in America, 1880–1950.* She interviewed the "witty, subversive" carto-journalist, then 90 years old, shortly before his death in 1993. Certainly, the wide range of maps he made for *Fortune* in the 1940s show him reveling in his amateur status. He tweaks professional geographers for their staunch commitment to the Mercator projection, the standard flat map of the world ("a dangerous map for world strategy," reads one caption). And he always finds space to explain, through clever illustrations and entertaining examples, exactly why he chose a particular style of mapmaking for a particular purpose. Never condescending, Harrison takes the tone of a fellow layman, and enthusiastic amateur, who just happened to discover what terms like "azimuthal equidistant" meant himself.

His map of "The World Divided," published in August 1941, centers on the North Pole, with the rest of the world a great circle around it. (In a typically charming sidebar, Harrison explains the principle behind this map with two cartoons of a dancer. In the first, she stands at rest, with her skirt as the globe; in the next, she's twirling, with the globe having risen and flattened to a disk.) This world, "divided" into Axis nations, Anti-Axis and various positions in between by color and shading, is also inextricably united: The U.S.S.R. and Alaska touch on this map, while the Aleutian Islands are part of one long, necklace-like chain that extends through Japan and down to the Philippines.

This view was prescient, of course. In March 1942, a nearly identical "polar equidistant" map appeared in *Fortune,* titled "One World, One War." Indeed, once the United States entered the war, Harrison's maps became integral to *Fortune*'s reports on U.S. action and changing strategic situations worldwide. The sweeping, slightly rounded views gave "a sense of direction, a sense of the movement of the war," says Schulten.

Harrison's work also serves to underscore the advances in air navigation at the time. As Schulten wrote in her book, "The use of a polar route to connect Japan to Alaska effectively transformed the Pacific from a massive body of water protecting the United States into a smallish lake."

In 1942, *Fortune* printed a stunning series of colorful, fold-out maps of the Pacific, Atlantic and Arctic arenas. These, says Perry, "make you see why something is happening in the world: Why are our troops in North Africa?... Why are your sons and neighbors dying in different places?" His vivid creations were phenomenally popular, reprinted by the military and various airlines and displayed at post offices.

Capitalizing on the cartographer's celebrity, *Fortune* published a book of Harrison's wartime maps in 1944. *Look at the World: The Fortune Atlas for World Strategy* was an instant best seller. Immediately after World War II, Harrison collaborated on numerous books like *Compass of the World: A Symposium on Political Geography* and *Maps and How to Understand Them,* an unusual hybrid textbook/political tract/advertising pamphlet published by Consolidated Vultee Aircraft Corporation promoting "Air Supremacy—For Enduring Peace."

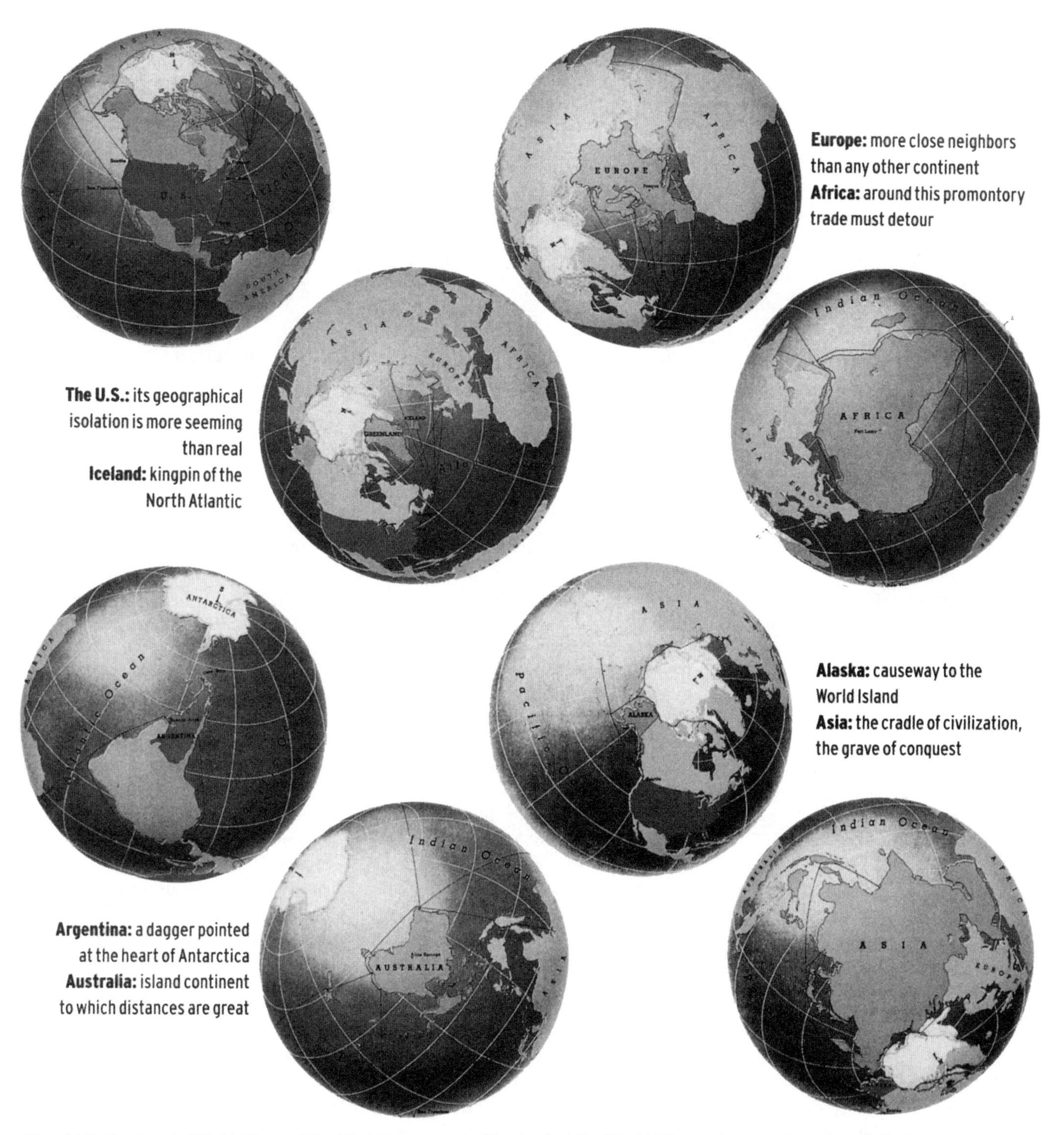

Planet Life Harrison's "Eight Views of the World" was created for *Look at the World.* The captions are from the original.

Today, though, *Fortune*'s "celebrated cartographer" is all but forgotten. The Mercator projection—that flat, 433-year-old map designed for an age of navigation, not aviation—has prevailed. "Perhaps his style of maps were really keyed into a time of crisis," Schulten speculates. "He had an advantage in the 1940s. While Rand McNally had to create maps that would last," Harrison, working for a monthly magazine, was making maps that were newsworthy. "He was drawing not just for the war, but for that week of the war."

We all know the fate of yesterday's news. It's too bad, though. Harrison's fresh-eyed perspectives and projections deserve a revival. Crisis or not, when it comes to viewing the world, we can always stand a "jolt… with a new and refreshing viewpoint."

Speaking of fortune, Navigator *is fortunate to have* ANN DE FOREST *as its resident cartographile.*

RESEGREGATION'S AFTERMATH

Mapping the stark difference in the quality of education between predominantly white and predominantly minority schools

Brad Holst

Fifty years ago in May, *Brown v. Board of Education* made segregated public education illegal. By 1990 the complexion of the nation's public schools had changed dramatically: individual schools more closely reflected the racial balance of the district in which they were located, and black students attended class with nearly twice as many white children, on average, as they had two decades previously.

But integration had already begun to lose momentum in the early 1970s. In fact, by some measures public schools have now been resegregating for more than a decade. Partly as a consequence, white children today have access to much better educational opportunities, on average, than do their minority counterparts.

Consider Los Angeles County. The accompanying map was generated by Jeannie Oakes, the director of the Institute for Democracy, Education, and Access at UCLA, using data she has collected on a range of "opportunity to learn" indicators. A white dot represents a high school with no significant problems. A gray dot identifies a school that has a serious problem with its teachers (fewer than 80 percent are certified by the state), its curriculum (fewer than two thirds of the available courses prepare students for a four-year college), or its facilities (buildings are so crowded that students must attend school in shifts, effectively shortening the academic year by seventeen days or more). The more areas in which a school has problems, the darker the dot.

A quick glance at the map reveals that problems are more common in schools located in predominantly black and Hispanic neighborhoods. The worst schools—the seven exhibiting serious problems with teachers, curricula, *and* facilities—collectively educate a population that is 94 percent black and Hispanic Meanwhile, not a single LA County school that is more than 90 percent white and Asian exhibits any serious problems with teachers, curricula, or facilities. Only 16 percent of the region's black and Hispanic students attend a school with none of the serious problems Oakes tracks; more than half (58 percent) of the region's white and Asian students do.

Numerous studies show that the caliber of teachers, the rigor of curricula, and the quality of facilities vary according to the race of the student body in public schools throughout the country. Teachers in predominantly black schools, for example, are less likely than teachers in predominantly white schools to have proper credentials or extensive teaching experience, to hold even a minor in the relevant subject, or to have attended a competitive college.

Not surprisingly, these differences matter. In L.A. County, graduation rates and college-preparedness rates for all races rise steadily with school quality. Indeed, both graduation rates and college-preparedness rates are twice as high in the best schools (those with no problems on Oakes's scales) as in the worst (those with problems in all three areas). Nationwide, 75 percent of white ninth-graders will graduate from high school in four years; barely half of black and Hispanic ninth-graders will. And even among those twelfth-grade black students who do graduate from high school, the average score on the standardized National Assessment of Educational Progress test is lower than the average score of white eighth-graders. This puts minority children on an unequal footing not only while they're in school but later in life as well.

A City of 2 Million Without a Map

SOMEWHERE IN this lakeside Central American town, there's a woman who lives beside a yellow car. But it's not her car. It's her address. If you were to write to her, this is where you would send the letter: "From where the Chinese restaurant used to be, two blocks down, half a block toward the lake, next door to the house where the yellow car is parked, Managua, Nicaragua."

Try squeezing that onto the back of a postcard. Come to that, try putting yourself in the place of the letter carriers who have to deliver such unruly epistles. How, for example, would they know where the Chinese restaurant used to be if it isn't there anymore? How would they know which way is "down," considering that "down," as employed by people in these parts, could as easily mean "up"?

How would they know which way the lake lies, when most of the time—in this topsy-turvy capital, punctured by the tall green craters of half a dozen ancient volcanoes—they cannot even see the lake? Finally, how would they know where the yellow car is parked, if its owner happens to be out for a spin?

Somehow, the people who live here have figured these things out. Granted, they've had practice. After all, most Managua street addresses take this cumbersome and inscrutable form. "We don't have a real street map," concedes Manuel Estrada Borge, vice president of the Nicaragua Chamber of Commerce, "so we have an amusing little system that no one from anywhere else can understand."

Welcome to Managua, quite possibly the only place on Earth where upward of 2 million people manage to live, work, and play—not to mention find their way around—in a city where the streets have no names.

No numbers, either. Well, that isn't quite true. A few Managua streets do indeed have conventional names. Some houses even have numbers. But no one hereabouts ever uses them. Why bother? Managuans have their own amusing little system to sort these matters out, a system that has the amusing little side-effect of driving most visitors crazy.

"For people who've just come here," says a long-time Canadian resident of the city, "there's no way on God's Earth that they'd know what you're talking about."

What Managuans are talking about, when all is said and done, is an earthquake that shattered this city three decades ago. Before that time, Managua was an urban conglomeration much like any other, at least in the sense that it had a recognizable center. It also had streets that ran east and west or north and south, and those streets not infrequently bore names. And numbers.

But then, on Dec. 23, 1972, the seismological fault lines that zigzag beneath Managua shifted and buckled, with horrific results. Upward of 20,000 people were killed in the quake, and the city was pretty much reduced to rubble. The catastrophe thoroughly disrupted the old grid pattern of Managua's streets, so the city's surviving residents were obliged to devise a new way of locating things. They started with a landmark—a certain tree, for example, or a pharmacy or a plaza or a soft-drink bottling plant—and they went from there.

Nowadays, for example, if you wished to visit the small Canadian Consulate in Managua, you would present yourself at the following address: *De Los Pipitos, dos cuadras abajo*. In English, this means: From Los Pipitos, two blocks down.

Any self-respecting inhabitant of Managua knows that "Los Pipitos" refers to a child-welfare agency whose headquarters are located a little south of the Tiscapa Lagoon. Managuans also know that *abajo*, in this context, does not mean "down" in a topographical sense. It means "west," because the sun goes down in the west. (By the same token, in Managua street talk, "*arriba*," or "up," means "east." *Al lago*, which literally means "to the lake," is how Managuans say "to the north." For some inexplicable reason, when they want to say "to the south," Managuans say "*al sur*," which means "to the south.")

Just to make a complicated process even more perplexing, Managuans, who normally use the metric system, will often give directions by employing an ancient Spanish unit of measurement called the *vara*. They will say, "From the little tree, two blocks to the south, 50 *varas* to the east." Visitors will therefore need to know how long a *vara* is (0.847 meters). They will also need to know that the "little tree" is no longer little. It is actually quite tall.

A few years ago, the Nicaraguan postal agency considered scrapping the jerry-rigged system of street addresses. But nothing came of the project. Besides, the scheme actually does seem to work. Nedelka Aguilar, for example, has learned that you merely have to have a little faith. Born in Nicaragua, she left as a young girl and spent most of her youth in southern Ontario. Now she lives in Managua once more. Shortly after her return four years ago, she arranged to visit a woman who dwelled at that outlandish address—"From where the Chinese restaurant used to be, two blocks down, half a block toward the lake, next door to the house where the yellow car is parked." By this time, Aguilar spoke the Managua dialect of street addresses well enough to take in the gist of this information. But what about that yellow car?

"I said to the woman, 'How will I find you if the yellow car isn't there?'" Aguilar smiles and shakes her head at the memory. "The woman laughed. She said, 'The yellow car is always there.'"

—OAKLAND ROSS, ***The Toronto Star*** (liberal), Toronto, Canada, April 21, 2002

From *World Press Review*, July 2002, pp. 44-45. Originally appeared in *The Toronto Sun*, "A City of 2 Million People and No Map," April 21, 2002.

UNIT 5

Population, Resources, and Socioeconomic Development

Unit Selections

33. **The Longest Journey**, The Economist
34. **Borders Beyond Control**, Jagdish Bhagwati
35. **China's Secret Plague**, Alice Park
36. **The Next Oil Frontier**, BusinessWeek
37. **Mexico: Was NAFTA Worth It?**, Geri Smith and Christina Lindblad
38. **Dry Spell**, Christopher Conte
39. **An Indian Paradox: Bumper Harvests and Rising Hunger**, Roger Thurow and Jay Solomon

Key Points to Consider

- Give examples of how economic development adversely affects the environment. How can such adverse effects be prevented?
- How do you feel about the occurrence of starvation in developing world regions?
- What might it be like to migrate from your home to another country?
- In what forms is colonialism present today?
- For how long are world systems sustainable?
- How can drought impact economic development?
- Discuss the positive and negative aspects of NAFTA. Choose a perspective: Mexico, Canada, or the U.S.

Links: www.dushkin.com/online/
These sites are annotated in the World Wide Web pages.

African Studies WWW (U.Penn)
http://www.sas.upenn.edu/African_Studies/AS.html

Geography and Socioeconomic Development
http://www.ksg.harvard.edu/cid/andes/Documents/Background%20Papers/Geography&Socioeconomic%20Development.pdf

Human Rights and Humanitarian Assistance
http://www.etown.edu/vl/humrts.html

Hypertext and Ethnography
http://www.umanitoba.ca/faculties/arts/anthropology/tutor/aaa_presentation.new.html

Research and Reference (Library of Congress)
http://lcweb.loc.gov/rr/

Space Research Institute
http://arc.iki.rssi.ru/eng/

World Population and Demographic Data
http://geography.about.com/cs/worldpopulation/

The final unit of this anthology includes discussions of several important problems facing humankind. Geographers are keenly aware of regional and global difficulties. It is hoped that their work with researchers from other academic disciplines and representatives of business and government will help bring about solutions to these serious problems.

Probably no single phenomenon has received as much attention in recent years as the so-called population explosion. World population continues to increase at unacceptably high rates. The problem is most severe in the less developed countries where, in some cases, populations are doubling in less than 20 years.

The human population of the world passed the 6 billion mark in 1999. It is anticipated that population increase will continue well into the twenty-first century, despite a slowing in the rate of population growth globally since the 1960s. The first and second articles in this section focus on issues of migration. The next article deals with China's "secret plague:" AIDS. A focus on the rise of U.S. oil interests in the Central Asian Region comes next. The next article provides a retrospective on the impact of NAFTA in Mexico. The article, "Dry Spell," argues for more proactive governmental responses to drought. The last article deals with a serious paradox: although there is more food produced globally each year, hunger persists for millions.

A survey of migration

The longest journey

Freeing migration could enrich humanity even more than freeing trade. But only if the social and political costs are contained, says Frances Cairncross

"WITH two friends I started a journey to Greece, the most horrendous of all journeys. It had all the details of a nightmare: barefoot walking in rough roads, risking death in the dark, police dogs hunting us, drinking water from the rain pools in the road and a rude awakening at gunpoint from the police under a bridge. My parents were terrified and decided that it would be better to pay someone to hide me in the back of a car."

This 16-year-old Albanian high-school drop-out, desperate to leave his impoverished country for the nirvana of clearing tables in an Athens restaurant, might equally well have been a Mexican heading for Texas or an Algerian youngster sneaking into France. He had the misfortune to be born on the wrong side of a line that now divides the world: the line between those whose passports allow them to move and settle reasonably freely across the richer world's borders, and those who can do so only hidden in the back of a truck, and with forged papers.

Tearing down that divide would be one of the fastest ways to boost global economic growth. The gap between labour's rewards in the poor world and the rich, even for something as menial as clearing tables, dwarfs the gap between the prices of traded goods from different parts of the world. The potential gains from liberalising migration therefore dwarf those from removing barriers to world trade. But those gains can be made only at great political cost. Countries rarely welcome strangers into their midst.

Everywhere, international migration has shot up the list of political concerns. The horror of September 11th has toughened America's approach to immigrants, especially students from Muslim countries, and blocked the agreement being negotiated with Mexico. In Europe, the far right has flourished in elections in Austria, Denmark and the Netherlands. In Australia, the plight of the *Tampa* and its human cargo made asylum a top issue last year.

Although many more immigrants arrive legally than hidden in trucks or boats, voters fret that governments have lost control of who enters their country. The result has been a string of measures to try to tighten and enforce immigration rules. But however much governments clamp down, both immigration and immigrants are here to stay. Powerful economic forces are at work. It is impossible to separate the globalisation of trade and capital from the global movement of people. Borders will leak; companies will want to be able to move staff; and liberal democracies will balk at introducing the draconian measures required to make controls truly watertight. If the European Union admits ten new members, it will eventually need to accept not just their goods but their workers too.

Technology also aids migration. The fall in transport costs has made it cheaper to risk a trip, and cheap international telephone calls allow Bulgarians in Spain to tip off their cousins back home that there are fruit-picking jobs available. The United States shares a long border with a developing country; Europe is a bus-ride from the former Soviet block and a boat-ride across the Mediterranean from the world's poorest continent. The rich economies create millions of jobs that the underemployed young in the poor world willingly fill. So demand and supply will constantly conspire to undermine even the most determined restrictions on immigration.

For would-be immigrants, the prize is huge. It may include a life free of danger and an escape from ubiquitous corruption, or the hope of a chance for their children. But mainly it comes in the form of an immense boost to earnings potential. James Smith of Rand, a Californian think-tank, is undertaking a longitudinal survey of recent immigrants to America. Those who get the famous green card, allowing them to work and stay indefinitely, are being asked what they earned before and after. "They gain on average $20,000 a year, or $300,000 over a lifetime in net-present-value terms," he reports. "Not many things you do in your life have such an effect."

Such a prize explains not only why the potential gains from liberalising immigration are so great. It explains, too, why so many people try so hard to come—and why immigration is so difficult to control. The rewards to the successful immigrant are often so large, and the penalties for failure so devastating, that they create a huge temptation to take risks, to bend the rules and to lie. That, inevitably, adds to the hostility felt by many rich-world voters.

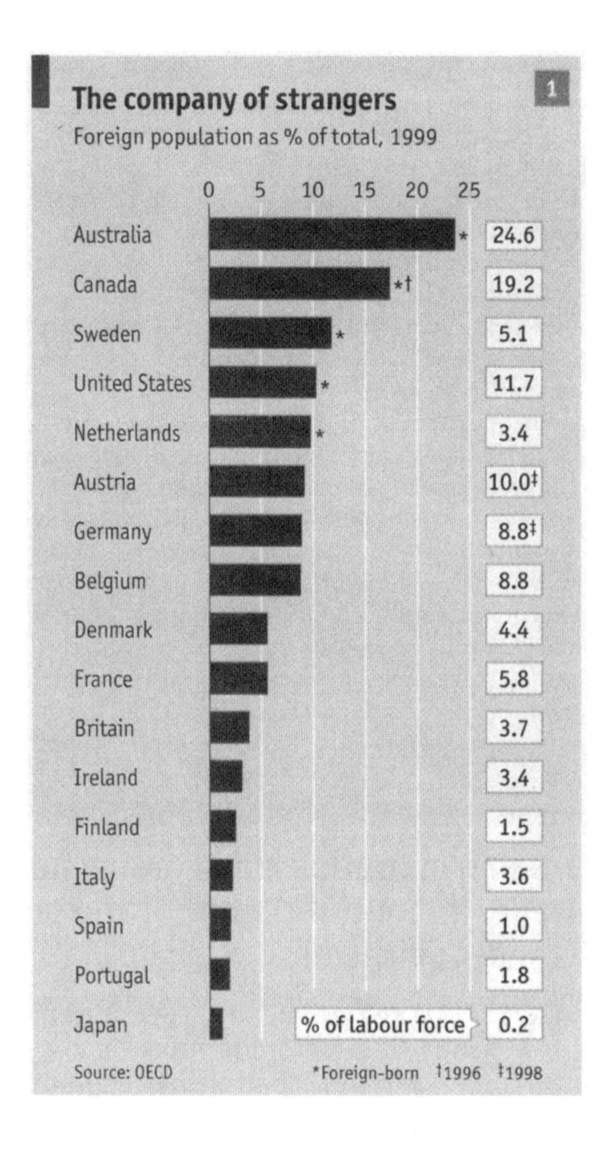

This hostility is milder in the four countries—the United States, Canada, Australia and New Zealand—that are built on immigration. On the whole, their people accept that a well-managed flow of eager newcomers adds to economic strength and cultural interest. When your ancestors arrived penniless to better themselves, it is hard to object when others want to follow. In Europe and Japan, immigration is new, or feels new, and societies are older and less receptive to change.

Even so, a growing number of European governments now accept that there is an economic case for immigration. This striking change is apparent even in Germany, which has recently been receiving more foreigners, relative to the size of its population, than has America. Last year, a commission headed by a leading politician, Rita Süssmuth, began its report with the revolutionary words: "Germany needs immigrants." Recent legislation based on the report (and hotly attacked by the opposition) streamlines entry procedures.

But there is a gulf between merely accepting the economic case and delighting in the social transformation that immigrants create. Immigrants bring new customs, new foods, new ideas, new ways of doing things. Does that make towns more interesting or more threatening? They enhance baseball and football teams, give a new twang to popular music and open new businesses. Some immigrants transform drifting institutions, as Mexicans have done with American Catholicism, according to Gregory Rodriguez, a Latino journalist in Los Angeles. And some commit disproportionate numbers of crimes.

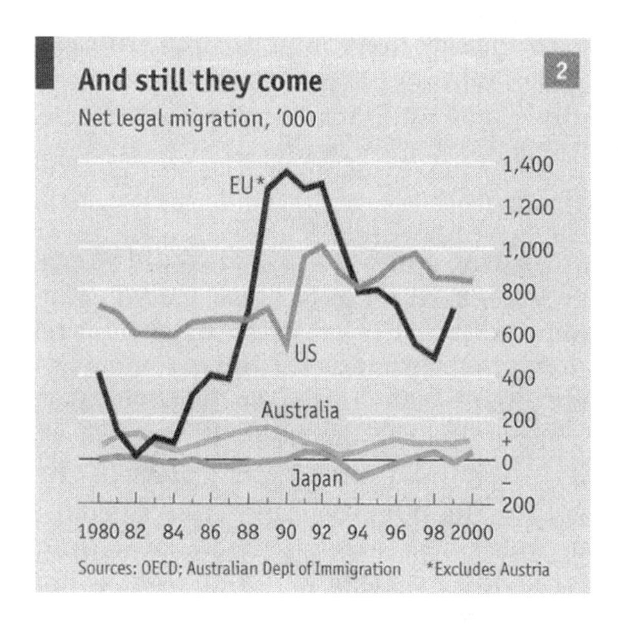

They also profoundly test a country's sense of itself, forcing people to define what they value. That is especially true in Europe, where many incomers are Muslims. America's 1.2m–1.5m or so Muslim immigrants tend to be better educated and wealthier than Americans in general. Many are Iranians, who fled extremist Islam. By contrast, some of the children of Germany's Turks, Britain's Pakistanis and France's North Africans seem more attracted to fundamentalism than their parents are. If Muslims take their austere religion seriously, is that deplorable or admirable? If Islam constrains women and attacks homosexuality, what are the boundaries to freedom of speech and religion? Even societies that feel at ease with change will find such questions hard.

No but, maybe yes

Immigration poses two main challenges for the rich world's governments. One is how to manage the inflow of migrants; the other, how to integrate those who are already there.

Whom, for example, to allow in? Already, many governments have realised that the market for top talent is global and competitive. Led by Canada and Australia, they are redesigning migration policies not just to admit, but actively to attract highly skilled immigrants. Germany, for instance, tentatively introduced a green card of its own two years ago for information-technology staff—only to find that a mere 12,000 of the available 20,000 visas were taken up. "Given the higher wages and warmer welcome, no Indians in their right minds would rather go to Germany than to the United States," scoffs Susan Martin, an immigration expert at Georgetown University in Washington, D.C.

Whereas the case for attracting the highly skilled is fast becoming conventional wisdom, a thornier issue is what to do about the unskilled. Because the difference in earnings is greatest in this sector, migration of the unskilled delivers the largest global economic gains. Moreover, wealthy, well-educated, ageing economies create lots of jobs for which their own workers have little appetite.

So immigrants tend to cluster at the upper and lower ends of the skill spectrum. Immigrants either have university degrees or no high-school education. Mr Smith's survey makes the point: among immigrants to America, the proportion with a postgraduate education, at 21%, is almost three times as high as in the native population; equally, the proportion with less than nine years of schooling, at 20%, is more than three times as high as that of the native-born (and probably higher still among illegal Mexican immigrants).

All this means that some immigrants do far better than others. The unskilled are the problem. Research by George Borjas, a Harvard University professor whose parents were unskilled Cuban immigrants, has drawn attention to the fact that the unskilled account for a growing proportion of America's foreign-born. (The same is probably true of Europe's.) Newcomers without high-school education not only drag down the wages of the poorest Americans (some of whom are themselves recent immigrants); their children are also disproportionately likely to fail at school.

These youngsters are there to stay. "The toothpaste is out of the tube," says Mark Krikorian, executive director of the Centre for Immigration Studies, a think-tank in Washington, D.C. And their numbers will grow. Because the rich world's women spurn motherhood, immigrants give birth to many of the rich world's babies. Foreign mothers account for one birth in five in Switzerland and one in eight in Germany and Britain. If these children grow up underprivileged and undereducated, they will create a new underclass that may take many years to emerge from poverty.

For Europe, immigration creates particular problems. Europe needs it even more than the United States because the continent is ageing faster than any other region. Immigration is not a permanent cure (immigrants grow old too), but it will buy time. And migration can "grease the wheels" of Europe's sclerotic labour markets, argues Tito Boeri in a report for the Fondazione Rodolfo Debenedetti, published in July. However, thanks to the generosity of Europe's welfare states, migration is also a sort of tax on immobile labour. And the more immobile Europeans are—the older, the less educated—the more xenophobic they are too.

The barriers need to be dismantled with honesty and care. It is no accident that they began to go up when universal suffrage was introduced. Poor voters know that immigration threatens their living standards. And as long as voters believe that immigration is out of control, they will oppose it. Governments must persuade them that it is being managed in their interests. This survey will suggest some ways in which that might be done.

Irresistible attraction

Who moves, and why

LEAVING one's home to settle in a foreign land requires courage or desperation. No wonder only a tiny fraction of humanity does so. Most migration takes place within countries, not between them, part of the great procession of people from country to town and from agriculture to industry. International migrants, defined as people who have lived outside their homeland for a year or more, account for under 3% of the world's population: a total, in 2000, of maybe 150m people, or rather less than the population of Brazil. Many more people—a much faster-growing group—move temporarily: to study, as tourists, or to work abroad under some special scheme for a while. However, the 1990s saw rapid growth in immigration almost everywhere, and because population growth is slowing sharply in many countries, immigrants and their children account for a rising share of it.

Counting migrants is horrendously difficult, even when they are legal. Definitions vary. Some countries keep population registers, others do not. The visitor who comes for a holiday may stay (legally or illegally) to work. Counting those who come is hard, and only Australia and New Zealand rigorously try to count those who leave. So nobody knows whether the rejected asylum- seeker or the illegal who has been told to leave has gone or stayed. But the overall picture is one of continuing growth in the late 1990s.

Between 1989 and 1998, gross flows of immigrants into America and into Europe (from outside the EU) were similar, relative to population size. About 1m people a year enter America legally, and some 500,000 illegally; about 1.2m a year enter the EU legally, and perhaps 500,000 illegally. In both America and Europe, immigration has become the main driver of population growth. In some places, the effects are dramatic. Some 36% of New York's present population is foreign-born, says Andrew Beveridge, a sociology professor at Queens College, New York: "It hasn't been that high since 1910," the last peak.

America at least thinks of itself as an immigrant land. But for many European countries the surge of arrivals in the 1990s came as a shock. For example, the Greek census of 2001 found that, of the 1m rise in the population in the previous decade (to 11m), only 40,000 was due to natural increase. "In a decade, Greece has jumped from being one of the world's least immigrant-dense countries to being nearly as immigrant-dense as the United States," notes Demetrios Papademetriou, co-director of the newly created Migration Policy Institute in Washington, D.C.

Asia too saw a burst of immigration in the 1990s, propelled initially by the region's economic boom. Foreign workers accounted for an increasing share of the growth in the labour supply in the decade to the mid-1990s. Chris Manning, an economist at the Australian National University in Canberra, reckons that foreign workers made up more than half the growth in the less-skilled labour force in Malaysia, perhaps one-third of the growth in Thailand and 15–20% of the growth in Japan, South Korea and Taiwan.

What makes all these people move? In the past, governments often imported them. In Europe, migration in the 1950s and 1960s was by invitation: Britain's West Indians and Asians, for example, first came at the government's request. Britain's worst racial problems descend from the planned import of textile and industrial workers to northern England. Now the market lures the incomers, which may produce less disastrous results.

Three forces often combine to drive people abroad. The most powerful is the hope of economic gain. Alone, though, that may not be enough: a failing state, as in Somalia, Sri Lanka, Iraq or Afghanistan, also creates a powerful incentive to leave. Lastly, a network of friends and relatives lowers the barriers to migrating. Britain has many Bangladeshi immigrants, but most come from the single rural district of Sylhet. Many host countries "specialise" in importing people from particular areas: in Portugal, Brazilians account for 11% of foreigners settling there; in France, Moroccans and Algerians together make up 30% of incomers; and in Canada, the Chinese share of immigrants is more than 15%.

Most very poor countries send few people abroad. Immigration seems to start in earnest with the onset of industrialisation. It costs money to travel, and factory jobs provide it. That pattern emerges strikingly from a study by Frank Pieke of the Oxford University of emigration from China's Fujian province. He describes how internal and overseas migration are intertwined. Typically, a woman from a family will go to work in a factory in a nearby province, supporting a man who then goes abroad and probably needs a few months to find himself a job.

Incentives to go, incentives to stay

Net immigration flows continue as long as there is a wide gap in income per head between sending and receiving countries. Calculations by the OECD for 1997 looked at GDP per head, adjusted for purchasing power, in the countries that sent immigrants to its rich members, and compared that figure with GDP per head in the host country. In all but one of its seven largest members, average annual income per person in the sending countries was less than half that of the host country.

Migrant flows peter out as incomes in sending and host country converge. Philip Martin, an economist from the University of California at Davis, talks of a "migration hump": emigration first rises in line with GDP per head and then begins to fall. Migration patterns in southern Europe in the 1980s suggested that the turning point at that time came at just under $4,000 a head. In a study for the European Commission last year of the prospective labour-market effect of EU enlargement, Herbert Brücker, of Berlin's German Institute for Economic Research (DIW), estimated that initially 335,000 people from the new members might move west each year, but that after ten years the flow would drop below 150,000 as incomes converged and the most footloose had gone. Net labour migration usually ends long before wages equalise in sending and host countries.

Migrants do not necessarily come to stay. They may want to work or study for a few months or years and then go home. But perversely, they are more likely to remain if they think that it will be hard to get back once they have left. "If you are very strict, you have more illegals," observes Germany's Ms. Süssmuth.

There has always been a return flow of migrants, even when going home meant a perilous return crossing of the Atlantic. According to Dan Griswold of the Cato Institute, a right-of-centre American think-tank, even in the first decade of the 20th century 20–30% of migrants eventually went home. And where migrants are free to come and go, many do not come in the first place. There is no significant net migration between the United States and Puerto Rico, despite free movement of labour. "It's expensive to be underemployed in America," explains Mr. Griswold. But in Europe, with its safety net of welfare benefits, the incentives to have a go are greater.

Tougher border controls deter immigrants from returning home. A book co-authored by Douglas Massey of the University of Pennsylvania, "Beyond Smoke and Mirrors", describes how in the early 1960s the end of a programme to allow Mexicans to work temporarily in America led to a sharp rise in illegal immigrants. Another recent study, published in *Population and Development Re-*

The going rate 3

Payments to traffickers for selected migration routes, $ per person

Route	$ per person	Route	$ per person
Kurdistan-Germany	3,000	North Africa-Spain	2,000-3,500
China-Europe	10,000-15,000	Iraq-Europe	4,100-5,000
China-New York	35,000	Middle East-US	1,000-15,000
Pakistan/India-US	25,000	Mexico-Los Angeles	200-400
Arab states-UAE	2,000-3,000	Philippines-Malaysia/Indonesia	3,500

Source: "Migrant Trafficking and Human Smuggling in Europe", International Organisation for Migration, 2000

view, also links tighter enforcement to a switch from temporary to permanent migration. Its author, Wayne Cornelius, says the fees paid to *coyotes*, people who smuggle migrants, have risen sharply. He found that, when the median cost of a *coyote*'s services was $237, 50% of male Mexican migrants went home after two years in the United States; but when it had risen to $711, only 38% went back. And, whereas the cost of getting in has risen (as have the numbers who die in the attempt), the cost of staying put has declined, because workplace inspections to catch illegals have almost ceased. The chance of being caught once in the country is a mere 1–2% a year, Mr. Cornelius reckons. So "The current strategy of border enforcement is keeping more unauthorised migrants *in* the US than it is keeping *out*."

Tighter controls in Europe are probably creating similar incentives to stay rather than to commute or return. A complex, bureaucratic system designed to keep many willing workers away from eager employers is bound to breed corruption and distortion. And the way that rich countries select immigrants makes matters worse.

Borders Beyond Control

Jagdish Bhagwati

A DOOR THAT WILL NOT CLOSE

INTERNATIONAL MIGRATION lies close to the center of global problems that now seize the attention of politicians and intellectuals across the world. Take just a few recent examples.

- Prime Ministers Tony Blair of the United Kingdom and José María Aznar of Spain proposed at last year's European Council meeting in Seville that the European Union withdraw aid from countries that did not take effective steps to stem the flow of illegal emigrants to the EU. Blair's outspoken minister for development, Clare Short, described the proposal as "morally repugnant" and it died amid a storm of other protests.
- Australia received severe condemnation worldwide last summer when a special envoy of the UN high commissioner for human rights exposed the deplorable conditions in detention camps that held Afghan, Iranian, Iraqi, and Palestinian asylum seekers who had landed in Australia.
- Following the September 11 attacks in New York City and Washington, D.C., U.S. Attorney General John Ashcroft announced several new policies that rolled back protections enjoyed by immigrants. The American Civil Liberties Union (ACLU) and Human Rights Watch fought back. So did Islamic and Arab ethnic organizations. These groups employed lawsuits, public dissent, and congressional lobbying to secure a reversal of the worst excesses.
- *The Economist* ran in just six weeks two major stories describing the growing outflow of skilled citizens from less developed countries to developed countries seeking to attract such immigrants. The "brain drain" of the 1960s is striking again with enhanced vigor.

These examples and numerous others do not just underline the importance of migration issues today. More important, they show governments attempting to stem migration only to be forced into retreat and accommodation by factors such as civil-society activism and the politics of ethnicity. Paradoxically, the ability to control migration has shrunk as the desire to do so has increased. The reality is that borders are beyond control and little can be done to really cut down on immigration. The societies of developed countries will simply not allow it. The less developed countries also seem overwhelmed by forces propelling emigration. Thus, there must be a seismic shift in the way migration is addressed: governments must reorient their policies from attempting to curtail migration to coping and working with it to seek benefits for all.

The reality is that little can be done to really cut down on immigration.

To demonstrate effectively why and how this must be done, however, requires isolating key migration questions from the many other issues that attend the flows of humanity across national borders. Although some migrants move strictly between rich countries or between poor ones, the most compelling problems result from emigration from less developed to more developed countries. They arise in three areas. First, skilled workers are legally emigrating, temporarily or permanently, to rich countries. This phenomenon predominantly concerns the less developed countries that are losing skilled labor. Second, largely unskilled migrants are entering developed countries illegally and looking for work. Finally, there is the "involuntary" movement of people, whether skilled or unskilled, across borders to seek asylum. These latter two trends mostly concern the developed countries that want to bar illegal entry by the unskilled.

All three problems raise issues that derive from the fact that the flows cannot be effectively constrained and must instead be creatively accommodated. In designing such accommodation, it must be kept in mind that the illegal entry of asylum seekers and economic migrants often cannot be entirely separated. Frustrated economic migrants are known to turn occasionally to asylum as a way of getting in. The effective tightening of one form of immigrant entry will put pressure on another.

SOFTWARE ENGINEERS, NOT HUDDLED MASSES

LOOKING at the first problem, it appears that developed countries' appetite for skilled migrants has grown—just look at Silicon Valley's large supply of successful Indian and Taiwanese computer scientists and venture capitalists. The enhanced appetite for such professionals reflects the shift to a globalized economy in which countries compete for markets by creating and attracting technically skilled talent. Governments also perceive these workers to be more likely to assimilate quickly into their new societies.

This heightened demand is matched by a supply that is augmented for old reasons that have intensified over time. Less developed countries cannot offer modern professionals the economic rewards or the social conditions that they seek. Europe and the United States also offer opportunities for immigrant children's education and career prospects that are nonexistent at home.

These asymmetries of opportunity reveal themselves not just through cinema and television, but through the immediacy of experience. Increasingly, emigration occurs after study abroad. The number of foreign students at U.S. universities, for example, has grown dramatically; so has the number who stay on. In 1990, 62 percent of engineering doctorates in the United States were given to foreign-born students, mainly Asians. The figures are almost as high in mathematics, computer science, and the physical sciences. In economics, which at the graduate level is a fairly math-intensive subject, 54 percent of the Ph.D.'s awarded went to foreign students, according to a 1990 report of the American Economic Association.

Many of these students come from India, China, and South Korea. For example, India produces about 25,000 engineers annually. Of these, about 2,000 come from the Indian Institutes of Technology (IITs), which are modeled on MIT and the California Institute of Technology. Graduates of IITs accounted for 78 percent of U.S. engineering Ph.D.'s granted to Indians in 1990. And almost half of all Taiwanese awarded similar Ph.D.'s had previously attended two prestigious institutions: the National Taiwan University and the National Cheng Kung University. Even more telling, 65 percent of the Korean students who received science and engineering Ph.D.'s in the United States were graduates of Seoul National University. The numbers were almost as high for Beijing University and Tsinghua University, elite schools of the People's Republic of China.

These students, once graduated from American universities, often stay on in the United States. Not only is U.S. graduate education ranked highest in the world, but it also offers an easy way of immigrating. In fact, it has been estimated that more than 70 percent of newly minted, foreign-born Ph.D.'s remain in the United States, many becoming citizens eventually. Less developed countries can do little to restrict the numbers of those who stay on as immigrants. They will, particularly in a situation of high demand for their skills, find ways to escape any dragnet that their home country may devise. And the same difficulty applies, only a little less starkly, to countries trying to hold on to those citizens who have only domestic training but are offered better jobs abroad.

A realistic response requires abandoning the "brain drain" approach of trying to keep the highly skilled at home. More likely to succeed is a "diaspora" model, which integrates present and past citizens into a web of rights and obligations in the extended community defined with the home country as the center. The diaspora approach is superior from a human rights viewpoint because it builds on the right to emigrate, rather than trying to restrict it. And dual loyalty is increasingly judged to be acceptable rather than reprehensible. This option is also increasingly feasible. Nearly 30 countries now offer dual citizenship. Others are inching their way to similar options. Many less developed countries, such as Mexico and India, are in the process of granting citizens living abroad hitherto denied benefits such as the right to hold property and to vote via absentee ballot.

However, the diaspora approach is incomplete unless the benefits are balanced by some obligations, such as the taxation of citizens living abroad. The United States already employs this practice. This author first recommended this approach for developing countries during the 1960s, and the proposal has been revived today. Estimates made by the scholars Mihir Desai, Devesh Kapur, and John McHale demonstrate that even a slight tax on Indian nationals abroad would substantially raise Indian government revenues. The revenue potential is vast because the aggregate income of Indian-born residents in the United States is 10 percent of India's national income, even though such residents account for just 0.1 percent of the American population.

UNSTOPPABLE

THE MORE DEVELOPED COUNTRIES need to go through a similar dramatic shift in the way they respond to the influx of illegal economic immigrants and asylum seekers. Inducements or punishments for immigrants' countries of origin are not working to stem the flows, nor are stiffer border-control measures, sanctions on employers, or harsher penalties for the illegals themselves.

Three sets of factors are behind this. First, civil-society organizations, such as Human Rights Watch, the ACLU, and the International Rescue Committee, have proliferated and gained in prominence and influence. They provide a serious constraint on all forms of restrictive action. For example, it is impossible to incarcerate migrants caught crossing borders illegally without raising an outcry over humane treatment. So authorities generally send these people back across the border, with the result that they cross again and again until they finally get in.

More than 50 percent of illegals, however, now enter not by crossing the Rio Grande but by legal means, such as tourist visas, and then stay on illegally. Thus, enforcement has become more difficult without invading privacy through such measures as identity cards, which continue to draw strong protests from civil liberties groups. A notable example of both ineffectual policy and successful civil resistance is the 1986 Sanctuary movement that surfaced in response to evidence that U.S. authorities were returning desperate refugees from war-torn El Salvador and Guatemala to virtually certain death in their home countries. (They were turned back because they did not meet the internationally agreed upon definition for a refugee.) Sanctuary members, with the aid of hundreds of church groups, took the law into their own hands and organized an underground railroad to spirit endangered refugees to safe havens. Federal indictments and convictions followed, with five Sanctuary members given three- to five-year sentences. Yet, in response to a public outcry and an appeal from Senator Dennis DeConcini (D-Ariz.), the trial judge merely placed the defendants on probation.

Sanctions on employers, such as fines, do not fully work either. The General Accounting Office, during the debate over the 1986 immigration legislation that introduced employer sanctions, studied how they had worked in Switzerland and Germany. The measures there failed. Judges could not bring themselves to punish severely those employers whose violation consisted solely of giving jobs to illegal workers. The U.S. experience with employer sanctions has not been much different.

Finally, the sociology and politics of ethnicity also undercut enforcement efforts. Ethnic groups can provide protective cover to their members and allow illegals to disappear into their midst. The ultimate constraint, however, is political and results from expanding numbers. Fellow ethnics who are U.S. citizens, legal immigrants, or amnesty beneficiaries bring to bear growing political clout that precludes tough action against illegal immigrants. Nothing matters more than the vote in democratic societies. Thus the Bush administration, anxious to gain Hispanic votes, has embraced an amnesty confined solely to Mexican illegal immigrants, thereby discarding the principle of nondiscrimination enshrined in the 1965 Immigration and Nationality Act.

MINDING THE OPEN DOOR

IF IT IS NOT POSSIBLE to effectively restrict illegal immigration, then governments in the developed countries must turn to policies that will integrate migrants into their new homes in ways that will minimize the social costs and maximize the economic benefits. These policies should include children's education and grants of limited civic rights such as participation in school-board elections and parent-teacher associations. Governments should also assist immigrants in settling throughout a country, to avoid depressing wages in any one region. Greater development support should be extended to the illegal migrants' countries of origin to alleviate the poor economic conditions that propel emigration. And for the less developed countries, there is really no option but to shift toward a diaspora model.

Some nations will grasp this reality and creatively work with migrants and migration. Others will lag behind, still seeking restrictive measures to control and cut the level of migration. The future certainly belongs to the former. But to accelerate the progress of the laggards, new institutional architecture is needed at the international level. Because immigration restrictions are the flip side of sovereignty, there is no international organization today to oversee and monitor each nation's policies toward migrants, whether inward or outward bound.

The world badly needs enlightened immigration policies and best practices to be spread and codified. A World Migration Organization would begin to do that by juxtaposing each nation's entry, exit, and residence policies toward migrants, whether legal or illegal, economic or political, skilled or unskilled. Such a project is well worth putting at the center of policymakers' concerns.

JAGDISH BHAGWATI is University Professor at Columbia University and André Meyer Senior Fellow at the Council on Foreign Relations.

Reprinted by permission of *Foreign Affairs*, Vol. 83, No. 1, January/February 2003, pp. 98-104.

CHINA'S SECRET PLAGUE

How one U.S. scientist is struggling to help the government face up to an exploding AIDS crisis

BY **ALICE PARK**
Kunming

THEY LINE THE DUSTY ROADS OUTSIDE THE tiny villages of China's Henan province, several hours' drive from Beijing—mounds of dirt funneled into crudely shaped cones, like a phalanx of earthen bamboo hats. To the uninitiated, they look like a clever new way of turning over fields—an agricultural innovation, perhaps, meant to increase crop yields. But the locals know the truth. Buried under the pyramids, which now number in the thousands, are their mothers and fathers, brothers, sisters and cousins, all victims of AIDS. Like silent sentries, the dirt graves are a testament to China's worst-kept secret.

They are the reason Dr. David Ho has come to China. The New York City-based virologist was named TIME's 1996 Person of the Year for his pioneering work on the drug therapies that have largely quelled the AIDS epidemic in the U.S. and Europe. Now Ho is confronting the AIDS virus in its most populous stronghold. Up to 1 million Chinese are HIV positive, and that number could easily grow to 10 million by 2010, according to the Joint U.N. Program on AIDS. If current trends continue for another decade or so, China could overtake Africa, where 29 million people have been infected with the virus.

It's to head off that scenario that Ho has traveled more than a dozen times to China over the past three years, setting up labs, visiting clinics, gathering blood samples, educating health workers and negotiating the intricately layered bureaucracy of the Chinese health establishment. Ho's efforts—and those of other AIDS activists—finally paid off last week when, on World AIDS Day, the Chinese government took a lesson from its sluggish response to the severe acute respiratory syndrome (SARS) epidemic and launched its first big AIDS public-awareness campaign, complete with posters, TV spots and an unprecedented visit by Premier Wen Jiabao to a Beijing hospital, where he shook hands with AIDS patients.

TIME accompanied Ho and his team from the Aaron Diamond AIDS Research Center (ADARC) for two weeks earlier this year as he traveled from Kunming, the cosmopolitan capital of Yunnan province, where his drug-treatment and vaccine projects are based, to the remote border town of Ruili, where heavy heroin trafficking and a thriving sex trade create a perfect HIV breeding ground, to Beijing, for his meetings with party leaders, including the newly appointed Minister of Health, Wu Yi. Everywhere Ho went, his mission was the same: to persuade Chinese officials to step up their modest anti-AIDS efforts and commit the resources necessary to launch a comprehensive nationwide program, modeled on the projects he has begun in Yunnan.

Kunming, Yunnan

THE NEATLY DRESSED HUSBAND AND WIFE ARE IN THEIR 50S and comfortingly average looking. Their once-smooth dark skin is now veined and burnished to a proud sheen, reflecting the decades of hard work they have put into raising a family, earning their salaries and, now, battling HIV.

They seem out of place in the world of AIDS. Neither injects drugs. Neither has had any contact with the sex trade. But they represent the newest and most troubling front in China's war against the AIDS virus. As in other countries hit by HIV, the epidemic in China began in the margins of society—among migrant workers, drug users and prostitutes—and then gradually entered the mainstream population. In China this process was facilitated by the government, which, through the tragic mismanagement of its blood-buying program in the early 1990s, permitted blood-collecting practices that ended up contaminating the country's blood supply with HIV. Anyone who gave blood or received a transfusion during that period was at high risk of contracting the virus—and then passing it on to his or her partners during intercourse.

That was how this couple, who declined to give their names, got the AIDS virus. They have kept it a secret from everyone but their immediate family, preferring not to risk being ostracized by their community. "Nobody knows," says the wife quietly. "They would not understand." The husband, as far as they can determine, was the first to get infected, perhaps from blood transfusions during surgery. It wasn't until his wife required an operation in 2001, however, that they were both found to be HIV positive. "I could not believe it," she says. "I told them they were totally wrong, that their detection was wrong. I heard reports that there was HIV in China, but that was mainly from people who traveled overseas. We never thought the virus would get here, in our family."

In a way, they are the lucky ones. Along with 68 other patients, they are part of a treatment program that Ho established in Kunming. There they will get the latest antiretroviral medications and the same careful monitoring that AIDS patients in the U.S. receive, including regular measurements of their viral loads and their immune-cell counts and tests to determine how quickly the virus is mutating to resist the drugs.

The epidemic began in the margins of society—among migrant workers, drug users and prostitutes

The vast majority of the Chinese who are HIV positive have no such access and must make do with drugs that treat the side effects of the disease—antibiotics for mouth sores and pneumonia, creams for skin lesions. Others rely heavily on traditional Chinese herbal medicines, which have no documented record of success. And even for those who are able to squeeze into one of the small studies supported by foreign aid groups, there is no guarantee of receiving proper follow-up care. "We have heard of places in China where the drugs are delivered but there is no training of the doctors in how to use them," says Ho. "We stress to them that drug treatment for AIDS is not like food relief, where the food is just dropped off."

As powerful as the AIDS drugs are, HIV mutates so rapidly that if the antiretroviral compounds are not properly administered, they are quickly rendered useless not just for that patient but for every other patient exposed to the mutated virus. It's a concept that is difficult for even the best-intentioned patients here to appreciate. TIME spoke with a patient advocate, 31, who goes by the pseudonym Ke'Er. He was infected after selling blood and was admitted to a study in Beijing that provided free U.S. antiretroviral drugs, but he accidentally left his two-month supply on the train after his most recent visit to the city. "I dared not tell my doctor," he said, "because I felt bad that I was offered this opportunity but I lost my medicine. So I found a Thai drug cocktail that is similar, and I'm taking that now." He doesn't know what the Thai drugs are but was assured by a doctor in his village that they would help. Chances are they won't.

Even the best AIDS drugs properly administered can do only so much. What doctors really need to head off a runaway epidemic is an effective vaccine. In fact, it was a vaccine trial that took Ho to China in the first place. In a way, China is an ideal place to conduct vaccine research. Because it is home to huge numbers of people who are HIV negative but at high risk of developing AIDS, Ho will be able to inoculate some of them with his vaccine and find out whether they can generate an immune response robust enough to protect them in case of a future exposure to HIV.

He is scheduled to inoculate his first healthy volunteers in New York with the U.S. version of the vaccine this week. Before he can begin testing a vaccine in China, however, Ho needs to know more about the virus strains circulating there. To protect against HIV, any experimental AIDS vaccine must be designed to match the rapidly changing strains moving through a population. Ho needs access to the blood of a lot of HIV-positive patients, so when he started looking for a place in China to conduct his trials, he turned first to Yunnan, a province with one of the greatest numbers of HIV and AIDS cases. His hope was that health officials there, who see the daily toll the disease takes, would be more willing to accept help from an outsider. It wasn't that simple.

Tracking HIV

YUNNAN IS CHINA'S FOURTH LARGEST province and historically one of its most mysterious and remote. (Its picturesque landscape of verdant hills and rustic villages inspired the legend of Shangri-La.) Its distance from the political leaders in Beijing has traditionally made it something of an outlaw province, home to dozens of minority groups and, in centuries past, feudal warlords who ruled with nearly absolute control. Today it is the gateway for heroin traffic that drifts into China from Burma, Vietnam and Laos.

As many as 10 million Chinese could be infected by 2010

Scattered along the drug route is China's largest concentration of heroin addicts. Yunnan has the highest IV-drug-use rates in China, and a recent U.N. AIDS report estimates that anywhere from 50% to 80% of the users are carrying the AIDS virus. HIV spread via unprotected sex is also on the rise here, accounting for 15% of HIV infections in 2000. All told, say health officials in Yunnan, this single province accounts for one-third of China's reported AIDS cases.

Given the Chinese penchant for careful record keeping, it's no surprise that officials here have been collecting and analyzing information on these cases for more than a decade. But access to the data—especially for outsiders—has been carefully guarded. The man in charge of generating the statistics is Dr. Lu Lin, director of the Yunnan Center for Disease Control (CDC), who has been monitoring HIV infection among the highest-risk groups in nearly 50 sites around the province since 1991.

A former prison guard with hooded eyes and a buzz cut who, at over 6 ft. tall, towers over most Chinese, Lu might seem a tough nut to crack. But when Ho approached Lu and his colleagues three years ago with a proposal to collaborate on vaccine trials, Ho was surprised by the response he got. They were eager to cooperate, he recalls, but had little interest in a vaccine. They were more concerned with helping those already struggling with the disease. "We wanted to push the vaccine," says Ho, "and they wanted to get more treatment for patients, more trained people and better labs to take care of the patients."

So for the past two years, Ho has retreated from his vaccine agenda and set up the pilot drug-treatment program in Kunming. Using funding from both ADARC and private donors, he has also built a clinic, set up a virology lab capable of performing basic viral-load tests and put together a state-of-the-art immunology lab—all of which will eventually absorb the testing required for the future vaccine studies.

In return, Ho has asked for access to the blood samples—some 24,000—collected from HIV patients throughout the province over the years. The samples will give him critical information about which populations in Yunnan would be suitable as the first subjects for his vaccine trials. "We realized we needed a quid pro quo," he says.

As part of that exchange, Lu's CDC team shared with Ho, in the first presentation of its kind to anyone outside the Chinese government, the details of AIDS penetration in Yunnan. Last March Lu informed Ho that in a 2002 survey of high-risk populations, 43% of IV drug users had shared needles with others in the past month, and that among female sex workers, 89% were unaware of their risk of contracting HIV. A majority of sex workers, about 60%, reported inconsistent condom use. Since they have begun collecting data, says Lu, there has been a 25% to 30% increase in HIV cases among IV drug users in the province. The incidence of HIV infection among sex workers has also risen steadily.

It was what Ho suspected but could never confirm without the data. Clearly, the few programs that the Chinese had put in place—distributing condoms and educating people about the dangers of unprotected sex—were having little effect on the spread of HIV, and most of the population was still both misinformed and uninformed about how dangerous the virus is. "We all appreciate that the epidemic in China was bigger than our expectations," he says. "We found ourselves taking on issues beyond just our research agenda. We realized that with a few more partners, we could—and should—do more educating, treating and training of people about AIDS in China."

To broaden the scope of his efforts, Ho enlisted the support of the newly appointed director of the province's Bureau of Health, Chen Juemin. Chen, to Ho's relief, is intent on addressing the AIDS epidemic in his province and is eager to have Yunnan serve as a testing ground for programs that Minister of Health Wu in Beijing will consider for the rest of the country. "This situation will not just go away," Chen told TIME. "We probably lost a chance [of controlling AIDS] because we did not open up publicly about our HIV work in the early 1980s. We didn't realize then that the disease was so serious and could spread so fast."

Mangxi, Yunnan

THE LAB, SUCH AS IT IS, CONSISTS OF just three rooms squeezed into a four-story building deep in Yunnan's southwestern town of Mangxi. The building has no elevator, and the external stairwell is bathed in the steamy heat that washes the entire region. Inside, however, in stark contrast to its tropical-outpost surroundings, are a few jewels of the modern microbiology trade—a state-of-the-art freezer for storing blood samples and an enzyme-linked immunosorbent assay (ELISA), a machine for screening HIV that can identify specific antibodies to the virus.

The equipment, including a computer and fax machine, all donated by Ho, will enable Mangxi to share vital data with Kunming, 280 miles away, and with Ho's group in the U.S. Yunnan's first case of HIV infection was discovered in Mangxi in 1989. Presumably the virus has been circulating here the longest; being able to include patients from the region in his study will enable Ho to tell how quickly the virus is mutating and which strains should be part of his experimental inoculation.

In Mangxi, Ho's priority is to sign up subjects, not an easy task when many of the prospective candidates are IV drug users and live in remote, largely inaccessible villages without telephones or newspapers; in fact, few of them can even read. Local health officials conduct their prevention efforts the old-fashioned way—going family to family, teaching couples how to use condoms and warning the young about the dangers of sharing needles.

One likely source of research subjects is the drug-rehabilitation camps that are blossoming all over Yunnan. A drug user picked up by the police is often forced to serve a mandatory three-month sentence in a rehabilitation camp, where calisthenics, lectures and daily treatment with a Chinese version of methadone are supposed to curb the addict's habit. Up to 20% of the inmates, by the guards' rough estimates, are HIV positive; because they are registered by the police, they can be tracked after they leave the camps. Eventually Ho wants to find and monitor 500 HIV-negative patients in the Mangxi area who are at high risk of becoming infected. Merging information on how many in this population eventually become HIV positive with data from the urban residents of Kunming will help him measure how quickly the virus is spreading.

Henan province

CHUNG TO, FOUNDER AND DIRECTOR of the nonprofit Chi Heng Foundation in Hong Kong, is one of the few outsiders who has penetrated the state-imposed isolation of the so-called AIDS villages in central China. He is all too familiar with the plight of small children orphaned by the disease. On a recent visit to a village in Henan, he watched an 8-year-old boy taking his father out for a walk. The boy was pushing his father along in a creaky wooden cart. The man was dying of AIDS and had been confined to his bed for weeks, too weak to walk. His son suggested the cart, hoping that a little fresh air would energize his ailing parent. A few weeks later, the father was dead.

"It was an unforgettable scene," says To. Using his own funds and donations, To has been helping these children continue their schooling, giving them a chance to free themselves from the taint of having a parent—or both parents, in some cases—die of AIDS.

In heavily affected provinces like Henan, Hebei and Shaanxi, an entire generation is vanishing in the shadow of AIDS. In family after family, mothers and fathers are dying, leaving as many as 200,000 children in Henan alone either parentless or in the care of aging grandparents. Ho and his colleagues were the first foreign group officially allowed to visit one of its villages, Wenlou. At the local hospital, only two doctors care for more than 1,000 HIV-positive patients, and they were trained not by the Chinese health system but by one of Ho's colleagues based in China.

Here, unlike in Yunnan, HIV is spread not through illegal behavior but through blood donation. In the early 1990s, the Chinese leadership launched a blood drive and paid donors for their plasma. It was a program intended to benefit all Chinese—the poor by giving them a way to supplement their income, and the rest of China by replenishing the national blood banks' dangerously low stocks. "It was like a poverty-relief program," says a Henan resident who gave plasma in 1993 and became infected. Through campaigns in the villages and schools, the government encouraged rural farmers and factory workers to sell their plasma for 40 yuan ($5). The good intentions backfired when "bloodheads," as some of the unofficial blood collectors came to be known, found a way to extract more plasma from fewer donors. Those running some stations pooled and processed the blood. Then they sent the plasma, containing useful proteins, to the blood banks and reinjected red and white blood cells, which can house HIV, into the donors. This enabled people to give several times a day, and nobody seemed to realize how dangerous the practice was. Infected blood now flowed through hundreds of thousands of residents in the central provinces, shifting the epicenter of AIDS cases, many experts believe, from Yunnan to the heart of China.

Henan and its neighbors, Ho has decided, cannot wait for his program to become established in Yunnan. In his proposal to the Ministry of Health, Ho has modified his plan to include testing, treatment and prevention projects for Henan and Yunnan. "They desperately want help," he says of the doctors he met in Wenlou. "They obviously have the data on AIDS patients but are afraid to show us."

Even today, 1 out of 5 Chinese have never heard of AIDS

That fear is well founded. Adding to the stigma surrounding AIDS in these villages is the role that local leaders played in the blood-buying program. "Many government officials made a lot of money," says the patient advocate who calls himself Ke'Er. To protect themselves, they wrapped their villages in the cloak of state secrecy, effectively sealing off AIDS patients from foreign aid groups as well as health officials from other provinces. AIDS-care centers still won't put the word AIDS on their doors, opting instead for such intentionally obscure labels as "home garden."

To break through this barrier of fear, Ho has encouraged Health Minister Wu to visit the AIDS villages in Henan. Wu's visit would be the first by someone in her post and would send, Ho hopes, a powerful message that the government is more interested in controlling the epidemic than in assigning blame. Wu was appointed Health Minister when her predecessor, with whom Ho had begun his project, was fired by the Communist Party for mishandling the SARS outbreak—denying its existence until the epidemic was out of hand. "SARS was a big kick in the pants for China," Ho says. "They were tainted by the SARS experience, and the health officials there now want to do the right things with AIDS."

Ho doesn't expect miracles. Many of the cultural traditions that make it difficult for the Chinese people and their government to openly address a sexually transmitted disease are too deeply rooted for one man to change. A recent survey by Futures Group Europe and Horizon Research Group revealed that 20% of Chinese still have not heard of AIDS and that only 5% have had an HIV test. Ho is convinced that even if just part of his program is put in place, it will save lives. "If we had known how difficult the process was going to be, I'm not sure we would have embarked on it," he says, reflecting on his work of the past three years. "We put up with a lot. But as AIDS researchers, we could not continue to be distant from the vast majority of patients."

The work, after all, is just beginning. Ho's team in New York City has analyzed the first material from the blood samples. "It looks really good," says Ho, visibly brightening at the prospect of finally starting up his vaccine studies. "Any one of the sites in Yunnan would work well for a vaccine trial." Starting those trials will mean China is that much closer to controlling HIV and slowing the spread of those earthen graves of family members claimed by AIDS.

CENTRAL ASIA

THE NEXT OIL FRONTIER

America carves out a sphere of influence on Russia's borders

It's Happy Hour at Fisherman's Wharf, an expatriate hangout in Baku, a port on the Caspian Sea in the former Soviet republic of Azerbaijan. The place is just around the corner from the town's only McDonald's, and on a Friday in April, a gaggle of Brits, Americans, and Aussies are gathered on bar stools to munch peanuts, quaff beer, and shoot the bull. Talking about Web access in this authoritarian, Muslim country, one guy, looking as if he had just returned from a long stint on an offshore oil rig, says to his buddy: "Yeah, but can you get hustler.com?"

The rugged oil worker is a type Americans can readily identify. Most Americans, though, couldn't find Azerbaijan on the map. And they probably wouldn't be able to find—or spell—Kyrgyzstan, Uzbekistan, Kazakhstan, or Tajikistan. But American soldiers, oilmen, and diplomats are rapidly getting to know this remote corner of the world, the old underbelly of the Soviet Union. The game the Americans are playing has some of the highest stakes going. What they are attempting is nothing less than the biggest carve-out of a new U.S. sphere of influence since the U.S. became engaged in the Mideast 50 years ago. The result could be a commitment of decades that exposes America to the threat of countless wars and dangers. But this huge venture—call it an Accidental Empire—could also stabilize the fault line between the West and the Muslim world and reap fabulous energy wealth for the companies rich enough and determined enough to get it.

The buildup has been breathtakingly fast. Consider:

- A year ago, not a single U.S. soldier was in the region. Today, roughly 4,000 servicemen and women are building bases, assisting the Afghan war, and training anti-insurgency troops along a rim of peril stretching 2,000 miles from Kyrgyzstan, on China's border, to Georgia, on the Black Sea. In early May, U.S. advisers started training antiguerrilla forces in the Pankisi Gorge in Georgia, where Muslim insurgents believed to be connected to al Qaeda are taking refuge from their struggle against Russian troops in Chechnya. A few days before that, Defense Secretary Donald H. Rumsfeld declared on a visit to Kyrgyzstan, where the U.S. Air Force has a base, that coalition troops would stay there "as long as necessary."
- From incidental sums fewer than five years ago, the amount of U.S. investment in the region has jumped to $20 billion.
- The energy giants have revved up their commitment to the Caspian region, one of the world's last undeveloped clusters of fields. Major investors include ChevronTexaco Corp., Exxon Mobil Corp., BP PLC, and Halliburton. BP alone plans to plow as much as $12 billion into the region over the next eight years.
- U.S. government aid is on track to jump 50% from pre-September 11 levels, to $809 million a year.

Every day, Americans are digging themselves in deeper into this part of the world, where 74 million people bring an exotic mix of Turkic, Mongol, Persian, and Slavic influence. What is fast evolving is a policy focused on guns and oil. The guns are to protect the local regimes from Islamic radicals and provide a staging area for attacks on Afghanistan. The goal is "to get rid of terrorism, not just get it out of Afghanistan," says A. Elizabeth Jones, Assistant Secretary of State for European and Eurasian Affairs. The guns, of course, will also protect the oil—oil that Washington hopes will lessen the West's dependence on the Persian Gulf and also lift the nations of the Caucasus and Central Asia out of their grinding poverty. "If you have prosperity, you have stability," Jones says.

Estimates of the Caspian oil pool vary greatly—from 200 billion barrels, on the level of a Saudi Arabia, to fewer than 100 billion barrels, still on a par with the reserves of the North Sea and at current oil prices worth $2.7 trillion. The Caspian could have a huge impact on the ability of OPEC to influence the oil market, says a U.S. government energy analyst. By 2010, the Caspian could account for 3% of global oil output, according to Moscow brokerage Renaissance Capital.

ChevronTexaco was the pioneer: In 1993, it bought into the huge Tengiz field in Kazakhstan. In October, 2001, almost $4 billion in investment later, a Chevron-led consortium opened its 980-mile pipeline from Tengiz to the Russian port of Novorossisk on the Black Sea. BP's Caspian project is one of its biggest anywhere. ExxonMobil has stakes in the Tengiz field and in offshore deposits belonging to Kazakhstan and Azerbaijan. All three majors are hungry to get in on future finds. "I don't think ChevronTexaco's appetite for investment in this part of the world is satisfied yet," says Dennis Fahy, general manager of Chevron in Kazakhstan.

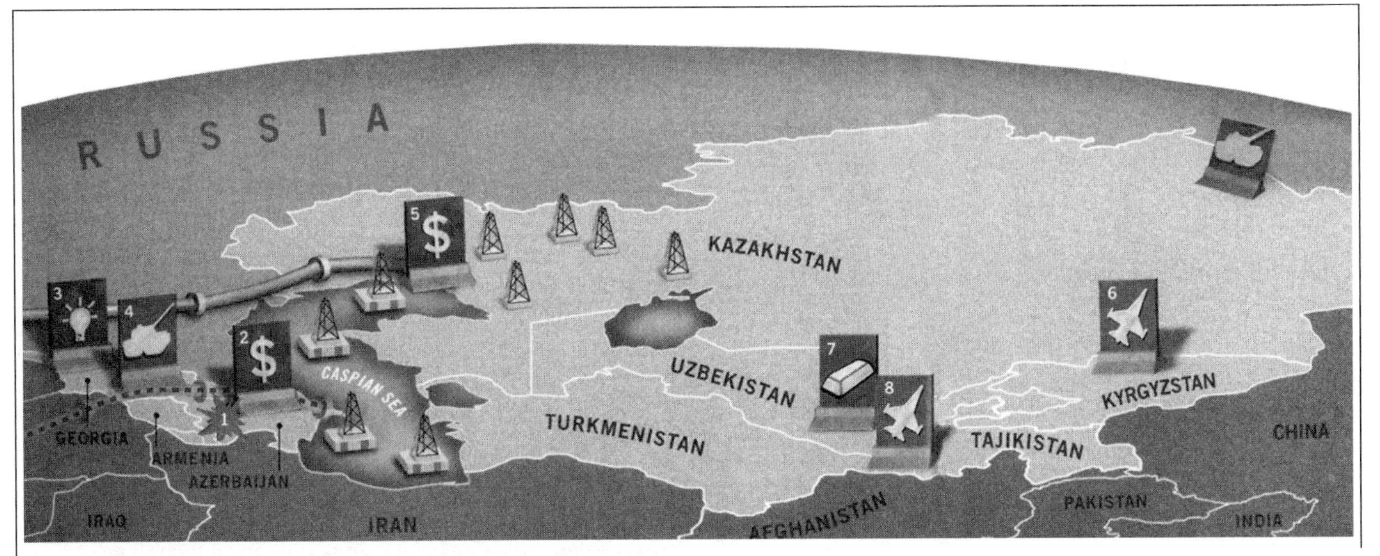

America's Game in Russia

Armenia

Population 3.8 million
GDP $2 billion
Natural Resources copper, gold
U.S. Objective 1 Deal with Azerbaijan over the disputed region of Nagorno-Karabakh

Azerbaijan

Population 8.1 million
GDP $5.7 billion
Natural Resources 0.9 billion tons of oil reserves
Investment Picture 2 BP will invest up to $12 billion in off-shore Caspian oil fields
U.S. Objective Thwart Iranian-sponsored militancy and strengthen Azeri Navy to secure oil investments

Georgia

Population 5.4 million
GDP $3.1 billion
Investment Picture 3 AES has invested $250 million in power
U.S. Objective 4 Use military to flush out militants

Kazakhstan

Population 14.9 million
GDP $20 billion
Natural Resources 1.82 trillion cubic meters of gas; 1.1 billion tons of oil
Investment Picture 5 ChevronTexaco, ExxonMobil, and others have invested $12.5 billion
U.S. Objective Oil security and access to ex-Soviet air bases

Kyrgyzstan

Population 4.7 million
GDP $1.4 billion
Natural Resources gold
Investment Picture: Procter & Gamble's sales last year grew by 40%
U.S. Objective 6 Troops stationed at ex-Soviet Manas air base for Afghanistan campaign

Tajikistan

Population 6.2 million
GDP $1.1 billion
U.S. Objective Thwart opium smuggling from Afghanistan

Turkmenistan

Population 5.4 million
GDP $3.3 billion
Natural Resources 2.83 trillion cubic meters of gas reserves

Uzbekistan

Population 25 million
GDP $6.5 billion
Natural Resources 1.85 trillion cubic meters of gas reserves, plus gold and uranium
Investment Picture 7 Newmont Mining plans more gold mines
U.S. Objective 8 Troops at Karshi air base are set for an Afghanistan campaign

Data: European Bank for Reconstruction & Development, U.S. Government, *BusinessWeek*

Key to the game are the pipelines, where diplomacy and oil-craft meet. The U.S. wants a pipeline that will help its friends in the region and freeze out its enemies—especially the Iranians, also located on the Caspian. That's why Washington is strongly discouraging plans by some oil majors to lay a pipeline across Iran, lobbying instead for a proposed $3 billion, 1,090-mile pipeline to carry up to 1 million barrels of oil a day from Baku through Georgia to the Mediterranean port of Ceyhan in NATO ally Turkey. BP, which is seeking to recruit other investors for the Baku-Ceyhan pipeline, is expected to make a final decision by June about going ahead. "Construction is going to be approved," says Richard Pegge, a senoir manager in *BP*'s Baku office.

THE SOUTHERN RIM: OIL-RICH, BUT STILL DIRT-POOR

	GDP PER CAPITA
SOUTHERN-RIM REPUBLICS	$586
CHINA	914
RUSSIA	1,790
IRAN	1,797
TURKEY	2,848
HUNGARY	4,757
U.S.	36,500

Data: *BusinessWeek*, Eurpean Bank for Reconstruction & Development, World Bank

Nothing is easy in this part of the world, however. Georgia has been wracked by civil war, organized crime, and terrorism. It's hardly a safe place for a pipeline. So the Pentagon is sending 150 military trainers to Georgia to help with anti-terrorism efforts and is helping Azerbaijan to bolster its Navy and modernize an air base for potential use by U.S. forces.

Not everyone is putting out the welcome mat. Russian hardliners see the southern rim thrust as U.S. encirclement. "Your foreign policy," a group of ex-military officers recently wrote President Vladimir V. Putin, is "the policy of licking the boots of the West."

Putin is trying to calm the hotheads. He may be calculating that his struggling country, barely able to supply its own armed forces, can benefit from the Pentagon's thrust. Putin and Bush plan to discuss U.S. military involvement in the Caucasus and Central Asia at their summit on May 24 in Moscow.

There's certainly plenty to talk about. On an April trip to the region, Defense Secretary Rumsfeld met with Kazakh President Nursultan A. Nazarbayev to discuss Pentagon access to local airfields. Some 1,000 troops of the U.S. Army's 10th Mountain Div. are already stationed at the ex-Soviet Khanabad Air Force Base in southern Uzbekistan. Fascinated by the female soldiers at the base, Uzbek guards offer to sell snapshots of women G.I.s riding motor scooters.

Russians are not the only ones nervous about U.S. troops in Central Asia. The State Dept.'s research shows that most people in Uzbekistan, Kazakhstan, and Kyrgyzstan oppose an extended U.S. military presence. "If the U.S. overstays its welcome in the region, it could alienate key allies in the war against terrorism," the department concluded in its Apr. 4 analysis. That risk also exists in oil-rich, BP-dominated Azerbaijan. "Bush sees us as the 51st state," scoffs Teymur Mamedov, a 32-year-old logistics manager for a Western oil-services company in Baku. "But it doesn't work that way. There's nothing to hold us together—only money, and that's not enough."

Then there's China, whose leaders suspect that the Pentagon's real goal is to keep an eye on, and if need be, contain China's activities in the region. Not to be outdone by the U.S., the Chinese are helping equip the Kazakh military.

The Chinese can play the power game, but in this chess match the U.S. has more pieces. Uzbek President Islam A. Karimov is grateful that the Pentagon-led campaign in Afghanistan dealt a blow to the local Islamic guerrilla group that fought alongside the Taliban. He's opening up the country's state-owned gold mines to $100 million in investment from Denver's Newmont Mining Corp.

For those expats who battle unyielding officials, impossible infrastructure and the sheer remoteness of it all, a stint in this part of the world can have its rewards. Even though most Baku residents lack properly filtered water, Western executives tied to the oil business are spending millions of dollars renovating 19th century townhouses with wrought-iron balconies as finely crafted as those in Paris. Most of the expat executives are middle-aged men, and with their fat wallets—let's face it, it's not their bulging waistlines—they are magnets for beautiful young local women. "Certainly, sexual harassment rules don't apply here," says one American male fortysomething businessman, recounting the perks of life in Baku.

Sensitive to the imperialism rap, the Bush Administration says its goal in the southern rim is to nurture prosperous, democratic societies. This is why the U.S. in mid-March inked an agreement with Uzbekistan. America pledged to protect the country from external threats in return for its pledge to liberalize its Soviet-style economy, improve its human-rights record, and ease government-imposed press censorship. In southern Azerbaijan, the State Dept. is funding a human-rights center in the town of Lenkoran, 25 miles from Iran. Still, even among center leaders, there's skepticism about America's purposes. "If there was no oil in Azerbaijan, I am sure America would not help us," says one of the staffers.

The Kremlin is sympathetic but not optimistic. "It was Russia's mission for so long to protect Western civilization from the Asians," says Vyacheslav A. Nikonov of the Polity Foundation, a Moscow political think tank. "If Americans are going to take over this job, God bless them."

Such sentiments aren't souring the American can-do spirit. James C. Cornell, president of RWE Nukem Inc. in Danbury, Conn., plans to double its uranium production in Uzbekistan. "When the U.S. is engaged militarily, it creates an umbrella for so many activities—not just business, but also education, culture," he says. "All things become possible." Trouble is, quagmires become possible, too.

By Paul Starobin, with Catherine Belton in Moscow, Stan Crock in Washington, and bureau reports

MEXICO: Was NAFTA Worth It?

A tale of what free trade can and cannot do

By Geri Smith and Cristina Lindblad

Piedad Urquiza probably doesn't know much about NAFTA, but she knows what it's like to have a steady job. Urquiza works at a Delphi Corp. auto-parts plant in Ciudad Juárez, just across the border from El Paso. The assembly line is a cross section of working-class Mexico, from twentysomethings raised in this border boomtown to veteran hands harking from the deep interior. In the years since NAFTA lowered trade and investment barriers, Delphi has significantly expanded its presence in the country. Today it employs 70,000 Mexicans, who every day receive up to 70 million U.S.-made components to assemble into parts. The wages are not princely by U.S. standards—an assembly line worker with two years' experience earns about $1.90 an hour. But that's triple Mexico's minimum wage, and Delphi jobs are among the most coveted in Juárez. "I like the environment, I like my colleagues," says Urquiza, a 56-year-old widow who assembles the switches that control turn signals. The daughter of a poor rancher, she dropped out of school after the seventh grade and has relied on her Delphi job to raise six children to adulthood—and, she hopes, to a better life.

The pact is one of the **BIGGEST**, most **RADICAL** trade experiments in history

Urquiza and millions of other Mexicans live out daily one of the most radical free-trade experiments in history. The North American Free Trade Agreement ranks on a par with Europe's creation of the euro and China's casting off Marxism for capitalism. It encompasses 421 million people and melds two first-world economies—the U.S. and Canada—with a struggling third-world country, Mexico. The bloc was seen as a bold attempt to demonstrate once and for all the free trade's vast power to turn a developing nation into a modern economy. If anything was a litmus test for globalization, NAFTA was it.

MEXICO Then & Now

Despite doubts among Mexicans, the benefits under NAFTA are numerous

1993	2003
Government	
Single-party dominated	Multiparty democracy
Gross domestic product	
$403 Billion	$594 Billion
Exports as % of GDP	
15%	30%
Oil as % of GDP	
18%	9%
Remittances by migrants in U.S.	
$2.4 Billion	$14 Billion

Data: Mexican Central bank, Economist Intelligence Unit

PROMISES, PROMISES

ON JAN. 1, NAFTA will celebrate its 10th anniversary. The assessment? The grand experiment worked in spades on many levels. American manufacturers, desperate for relief from Asian competition, flocked to Mexico to take advantage of wages that were a 10th of those in the U.S. Foreign investment flooded in, rising to an annual average of $12 billion a year over the past decade, three times what India takes in. Exports grew threefold, from $52 billion to $161 billion today. Mexico's per capita income rose 24%, to just over $4,000—which is roughly 10 times China's."NAFTA gave us a big push," Mexican President Vicente Fox told *BusinessWeek.* Fox notes proudly that Mexico's $594 billion economy is now the ninth-largest in the world, up from No. 15 a dozen years ago. "It gave us jobs. It gave us knowledge, experience, technological transfer."

Just as important, the pact spurred profound political change. Mexicans who backed open markets also wanted an open political system. Would the Institutional Revolutionary Party (PRI) have fallen from seven decades in power in 2000 if Mexico hadn't signed a treaty requiring government transparency, equal treatment for domestic and foreign investors, and international mediation of labor, environmental, and other disputes? It's hard to believe democracy would have come to Mexico as quickly without NAFTA.

Impressive milestones—and seemingly ample proof that free trade delivers the goods. But rightly or wrongly, a large proportion of Mexicans today believe the sacrifices exceeded the benefits. The Mexican mood is infecting other Latin countries, which after 15 years of gradually opening their own economies to trade and investment are showing pronounced fatigue with the "Washington consensus," the free-market formula preached by the U.S. and the International Monetary Fund. In an August poll of 17 Latin countries carried out by Chile-based Latinobarómetro, just 16% of respondents said they were satisfied with the way market economics were working in their countries. Thus NAFTA's perceived shortfalls are giving fresh ammunition to free trade's opponents. "Now you have a whole network of people organizing against the Free Trade Area of the Americas and globalization because of what has happened in Mexico under NAFTA," says Thea Lee, the AFL-CIO's chief expert on international trade pacts. That's an ironic twist: It was NAFTA, after all, that kicked the global free-trade movement into high gear, spurring forward the Uruguay round of global trade talks in the mid-1990s and setting the stage for China's entry into the World Trade Organization in 2001.

Why have so many Mexicans soured on NAFTA? One problem is that the deal was oversold by its sponsors as a near-magic way to turn Mexico into the next Korea or Taiwan. Ten years later, many think the pact has stopped paying dividends—and that Mexico has been unfairly neglected by a Washington consumed by the war on terror. Speaking before an audience of Mexican students on Nov. 11, Mexico's envoy to the U.N., Adolfo Aguilar Zinser, characterized NAFTA as "a weekend fling." The U.S., he said, "isn't interested in a relationship of equals with Mexico, but rather in a relationship of convenience and subordination." While Zinser's remarks cost him his job, his words struck a chord. In an October survey by a leading pollster, only 45% of Mexicans said NAFTA had benefited their economy. That's down from the 68% who in November, 1993, saw the pact as a strong plus. With the U.S. in a slump for the past three years, Mexicans are experiencing the downside of their close commercial ties with the colossus. Mexico's economy will grow by 1.5% this year, a poor showing for a developing country.

Mexico believed NAFTA would make it the U.S.'s top workshop. China got the job

In a larger sense, Mexicans feel shortchanged by globalization. They thought they would be America's biggest workshop. That honor now belongs to China, which this year surpassed Mexico as the U.S.'s No. 2 supplier. Mexican policymakers signed trade agreements with a total of 32 countries, and as a result consumers got cheaper and better goods. Yet local manufacturers of everything from toys to shoes, as well as farmers of

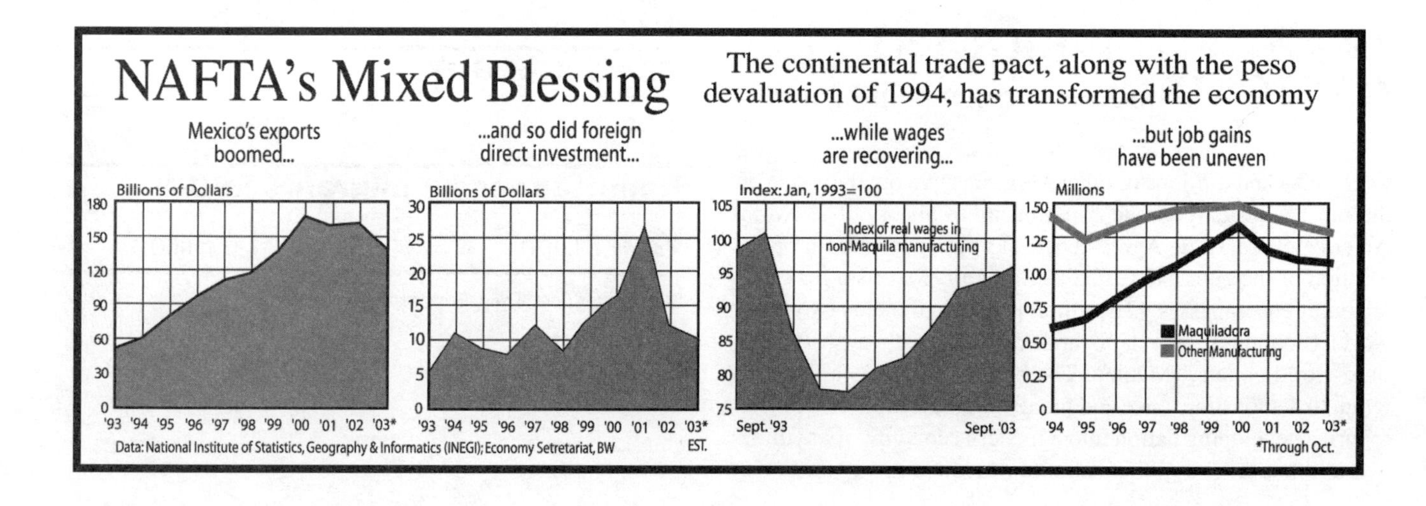

rice and corn, struggle to survive the onslaught of cheap imports. Mexicans hoped NAFTA would generate enough jobs to keep them at home. Instead, the jobless flock in ever-greater numbers across the border. Reforms that pressed on Mexico before NAFTA—modernizing the electricity sector, overhauling the tax code, shoring up the crumbling schools—are an even more difficult sell now that power is split among several parties.

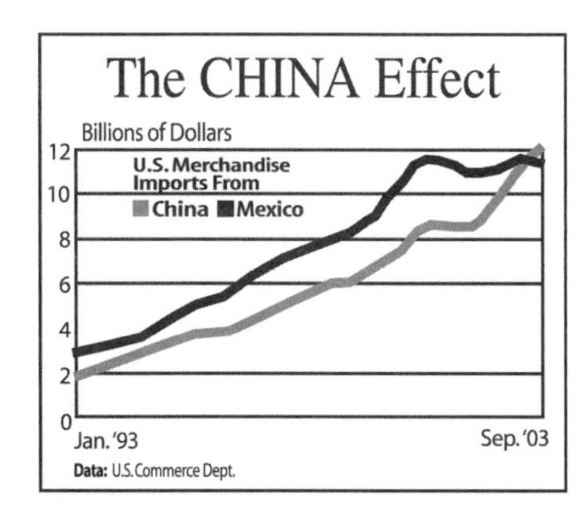

OPPORTUNITY KNOCKED

DO MEXICO'S woes disprove the value of free trade? Few would argue that NAFTA was a waste. "If we didn't have NAFTA, we'd be in far worse shape than we are today," says Andrés Rozental, president of the Mexican Council on Foreign Relations. If NAFTA has disappointed, it is in large part because the Mexican government has failed to capitalize on the immense opportunities it offered. "Trade doesn't educate people. It doesn't provide immunizations or health care," says Carla A. Hills, the chief U.S. negotiator in the NAFTA talks. "What it does is generate wealth so government can allocate the gains to things that are necessary." If a government doesn't allocate new wealth correctly, the advantages of free trade quickly erode. That is Mexico's plight. "NAFTA wasn't an end unto itself, but a means to something, and that something was precisely the need to go further in reform," says former Mexican President Carlos Salinas, one of NAFTA's principal architects. "It's like Alice in Wonderland—you have to run faster and faster if you want to stay in the same place. Globalization won't wait for you."

The outcome of Mexico's struggle to regain the momentum is of vital interest not just to Latin America but also to the U.S. The Bush Administration has made trade a vital part of its hemispheric agenda. Besides, the U.S. needs a stable, prosperous Mexico on its border to stem the flood of illegal immigration and drugs. Mexico's ability to get to the next stage will also show whether low-wage economies around the globe can hold their own against China. "Mexico cannot compete sewing brassieres and tennis shoes," says Roger Noriega, U.S. Under Secretary of State for the Western Hemisphere. "they cannot compete with China—who can? Mexico has to modernize so it can move forward."

NAFTA has already proven a powerful impetus to reform. Mexico did not hike its import tariffs when the peso crisis of 1994 hit. Encouraged, Washington stepped in with a $40 billion bailout package that helped Mexico stabilize its finances and return to the capital markets in just seven months. Although wrenching, the devaluation turbocharged NAFTA by dramatically lowering the costs of Mexican labor and exports. The government's fiscal discipline has earned the country a coveted investment-grade rating on its debt. And the current recession is mild by historic standards. Most analysts see growth quickening to 3.5% next year.

Yet even with a rebounding economy, Mexico will not generate enough jobs to accommodate its fast-growing workforce. While U.S. companies praise the work of their Mexican employees, they now make it abundantly clear that there are other, cheaper locales. An assembly line worker in Mexico earns $1.47 an hour; his counterpart in China makes 59¢ an hour, according to a new report by McKinsey & Co. Top Delphi executives have warned for months that some production may be shifted to China because of the many cost advantages it offers. "Delphi and other automotive suppliers are courted every day by other countries, not only with lower-cost labor but also with new incentives and tax breaks," says David B. Wohleen, president for electrical, electronics, safety & interior."Mexico will need to significantly pick up the pace to remain a competitive alternative," he warns.

No one feels the China threat more keenly than Daniel Romero, president of the National Council of the *Maquiladora* Export Industry. Mexico's *maquiladora,* which assembles goods for export using imported parts and components, had been around since the mid-1960s. Under NAFTA, the number of plants rose 67%, to 3,655 in seven years. Yet more than 850 factories have shut down since 2000, with many shifting to cheaper locales. Employment is down more than 20% from its peak of 1.3 million workers. Romero and a group of *maquiladora* managers traveled to China last year. They came away dispirited. "They have aggressive tax incentives, low salaries, very aggressive worker training, and a supply chain that allows them to have immediate access to the latest technology," says Romero.

The agriculture sector is suffering even more than the *maquiladoras*, as subsidized U.S. food imports flood the country. Some 1.3 million farm jobs have disappeared since 1993, according to a new report by the Carnegie Endowment for International Peace, a Washington think tank. "NAFTA has been a disaster for us," says Julián Aguilera, a pig farmer from Sonora. He and his peers have staged big demonstrations to protest a 726% increase in U.S. pork imports since the pact took effect. "Mexico was never prepared for this."

Nor was the U.S. As the *campesinos* lost their livelihood, they headed to the border. By most estimates, the number of Mexicans working illegally in the U.S. more than doubled, to 4.8 million between 1990 and 2000. Despite tightened security after September 11, hundreds of thousands continue to cross the border. The money sent back to their families will total $14 billion this year, more than the $10 billion Mexico expects in foreign direct investment.

The exodus has turned rural hamlets into ghost towns. Panindícuaro in Michoacán, one of Mexico's poorest states, has one the highest incidences of migration, with one out every seven people leaving. Panindícuaro's priest, Melesio Farías, recently held a funeral mass for a father in his mid-thirties who died trying to cross the Arizona desert. "I tell them to forget the U.S. and to work at home," says Farías. "But if Mexico can't offer them jobs, why should they?"

Salinas' band of technocrats and their successors didn't do enough to prepare vulnerable sectors for NAFTA'S onslaught. Long -promised programs to help 20 million *campesinos* switch to export crops never materialized. Nor has the government offered inducements to channel foreign investment into areas of the country where it is most needed. The six border states, along with the capital, nabbed 85% of foreign outlays year. Little has been done to foster a local supplier network for the import-dependent *maquiladoras*. Less than 3% of the industry's parts are sourced in Mexico. "Society at large and a good chunk of the economy have failed or refused to adjust to globalization," argues Luis Rubio, who heads the Center of Research for Development, a Mexico City think tank. "And the Mexican government has done absolutely nothing to help."

This laissez-faire attitude is in stark contrast to China. There, state-owned banks have bankrolled lavish investments in industrial parks, power plants, highways, and other infrastructure to provide low-cost facilities for foreign manufacturers. These multinationals had to source as many components as possible from domestic suppliers, and the government wasn't bashful about demanding transfers of technology to Chinese partners. Also, Beijing sealed off weak sectors like financial services or retailing. As a condition for entry into the WTO in 2001, China is phasing out these policies, but its domestic companies now have a head start.

Even if China-style tactics are not possible, Mexico could still hone its competitiveness. The PRI under Salinas took advantage of its monopoly on power to ram through painful reforms that paved the way for NAFTA. Now under a multiparty system, the politicians struggle to make difficult choices. Mexico will need to spend $50 billion to upgrade its power grid. But legislation to open the constitutionally protected sector to private investment has run aground on nationalist sentiment and union opposition, even while electricity rates are as much as 40% higher than in China. Grupo México, the world's third-largest copper producer, is considering moving refining operations to Amarillo, Tex., where electricity costs 4¢ per kilowatt-hour, vs. 8.5¢ in Sonora.

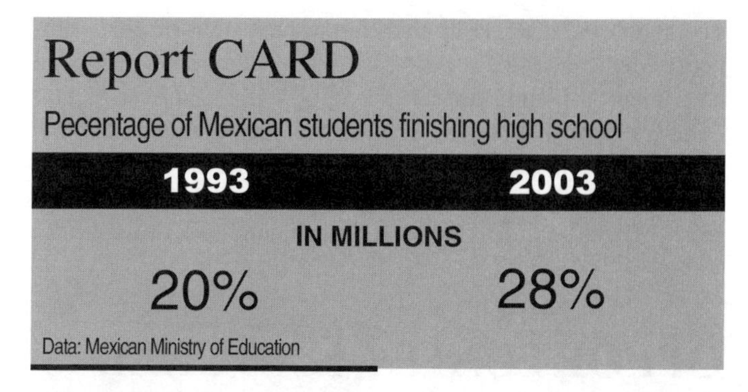

Education is another critical area where reform has stalled. William Spurr, head of the North American transport division of Canada's Bombardier, which builds railcars in Hidalgo, sees a need for more skilled workers. "There's a very good talent pool, but there aren't enough of them," he says."If I opened a plant in India, I'd have all the engineers and technicians I need." To be fair, the government's finances have been sapped by a $100 billion bank bailout after the peso crisis. Even under those circumstances, the number of science and engineering college grads has nearly doubled over the past decade, to 73,300. Yet that number still pales next to India, which graduated 314,000 students in those subjects, while China handed out diplomas to 363,000. Congress has so far foiled Fox's efforts to raise taxes to pay for improved education.

To get a glimpse of what the right training can do, consider the case of Tecnomec Agrícola, a maker of farm and earth-moving equipment based in Aguascalientes, in central Mexico. "We never had a tradition of exporting. NAFTA definitely changed that," says founder José Leoncio Valdés. It was hard going at first. "We couldn't get in to see people in the U.S. because we were from Mexico and they figured we were unreliable," recalls the 55-year-old engineer. Then in 2000, Valdés dispatched his son José to earn a degree in engineering and business administration at Massachusetts Institute of Technology. On his first spring break, young José conducted a weeklong session with Tecnomec managers. He used Lego blocks to build a replica of the factory and figure out how to better track inventory, boost quality, and control waste. Tecnomec soon boosted productivity by 21%. Now its exports total $1.4 million a year, nearly a quarter of the company's annual sales.

Mexico could use more Tecnomecs. Just 50 companies account for half of all of Mexico's exports—and the top tier is dominated by multinationals. Thousands of other Mexican businesses have gone under in the face of competition."We are at a watershed," says Jaime Serra Puche, Mexico's chief NAFTA negotiator. "Either we take the steps to become a true North American country or we just become a big Central American country."

Serra Puche is one of many prominent Mexicans trying to figure out how to improve NAFTA."If we were going to do it all over again today, I would insist on introducing a lot of considerations," says President Fox, who believes that NAFTA should be modeled more on the European Union, with provisions for free movement of labor and cross-border grants to compensate poorer countries for the dislocations. Proposals for a single currency, a North American energy cooperation plan have also surfaced. But don't expect any breakthroughs soon—not while the U.S. heads into elections and trade has re-emerged as a contentious issue.

So for now the burden will remain on Mexico. Salvador Kalifa, an independent economist based in Monterrey, recalls that when Spanish conqueror Hernán Cortés reached Mexico, he burned his boats to prevent crew members from fleeing. "With NAFTA, we burned our boats and threw ourselves into globalization," says Kalifa."There is no turning back."

DRY SPELL

Places can't stop drought from coming their way, but they can control its devastating effects.

BY CHRISTOPHER CONTE

Water officials in Denver didn't worry last April when the usual spring showers failed to materialize and people started watering their lawns two months ahead of schedule. After all, the city's reservoirs were more than three-quarters full, and one look at the snow-capped peaks of the Rocky Mountains seemed to promise a healthy spring run-off.

What they failed to check was the soil. Underneath the snow pack, it was seriously parched, and the mountain forests were desperately thirsty. When the spring melt began, the soil and trees drank up much of the water that would have filled streams in a normal year. May brought more trouble in the form of dry winds that vaporized much of the snow before it could melt, shrinking the snow pack to 13 percent of normal. Still, the water utility, blithely assuming reservoirs would refill as normal, merely recommended that people voluntarily cut their water use by 10 percent. Their light-hearted slogans—"Real Men Dry Shave" and "No Water, No Bikinis"—failed to stir much public concern.

It was only in July, when the flow of the south Platte River had slowed to less than 500 acre-feet of water compared with 30,000 normally and reservoirs had dropped to 60 percent of capacity, that Denver Water realized it had a crisis on its hands. In fact, it was confronting—and continues to confront—the worst drought in the central Rockies in 300 years.

The Mile High City is not alone in its distress. Normally, about 10 percent of the country suffers very serious drought at any time, but last summer the rate soared to 38 percent, and more than half of the country experienced abnormally dry conditions, if not outright drought. No region was immune. Some 18,000 private wells went dry in Maine. South Carolina, which went through its fifth consecutive year of dryness, was staggered by $520 million in timber losses due to slower tree growth and the loss of drought-weakened trees to pine beetles. Low river levels led to a surge in hydroelectric power prices in the usually rain-drenched Pacific Northwest. And in the Plains states, Kansas farmers last year reported $1 billion in crop losses due to lack of rain.

'Water is such a touchy subject here,' says Colorado's Brad Lundahl.

This year may not be much better. Agricultural analysts predict that the mild, dry winter in the heartland will lead to a plague of grasshoppers in every state west of the Mississippi River this summer. And the U.S. Army Corps of Engineers warned it might have to close this year's shipping season on the Mississippi and Missouri Rivers early because water levels may be too low for many barges.

While the recent drought has been, by many measures, as bad or worse than any in the past century, it isn't the first wake-up call to states that they can't afford to be lax about plans to deal with parched conditions. And there will be more warnings in the future. As the population continues to grow and shift both from rural areas to urban areas and from the humid East to the arid Southwest, water supplies are growing tighter. That suggests it may not take much of a deficiency in rainfall or snowfall to produce water shortages in the future. Add the possibility—unproven, but feared by many scientists—that global warming will lead to more frequent droughts, and you have a strong case that states—and localities—should get a lot more serious about dealing with drought.

Coming up dry

To their credit, states are becoming more aware of the risks. While just two states had formal drought plans in 1982, some 33 have such plans today, and seven more are working on the issue. But these plans generally deal with how governments will respond to a drought once its effects become apparent—an approach that falls far short of what hydrologists say is needed to avoid most of the damage. "If you wait until after a drought begins, all you can do is try to manage resources day to day," says Donald Wilhite, director of the National Drought Mitigation Center in Lincoln, Nebraska. "The time to prepare for a drought is before the drought begins."

Wilhite believes states have been shortsighted partly because the federal government is so quick to provide disaster-relief funds. That, in turn, diminishes the incentive to invest in preventive measures, such as monitoring conditions to detect when droughts might occur. "We have spent a lot of money putting out brushfires, and long-term monitoring has gotten the axe," says Barry Norris, the chief drought official for Oregon's Water Resources Department. For a capital investment of just $1.5 million, plus about $200,000 in annual operating expenses, Norris says he could install an effective network of equipment to monitor stream flows, snow pack and soil moisture levels. But in years when there is no drought, the expenditure for such a program can't compete with other state spending priorities. And when droughts occur, policy makers are too focused on short-term solutions to invest in policies that would pay off only in the long run.

Compounding the problem, state governments aren't set up to address drought issues comprehensively. In many cases, drought policy is the responsibility of agriculture departments, which in turn are beholden to farm constituencies that have grown adept at winning disaster-relief funds. In others, drought planning falls under the purview of emergency-preparedness agencies, which lack expertise in long-term water issues. And some state governments, especially in the West, feel thwarted by time-encrusted water-rights laws and powerful interest groups. "We could do more to be prepared for when the big droughts hit, but it's hard for me to get anything going," says Brad Lundahl, who as section manager for the Conservation and Drought Planning Section of the Colorado Water Conservation Board, is the state of Colorado's top drought official. "Water is such a touchy subject here, and the water community doesn't want to give us much authority."

The last straw

Denver's experience is a textbook example of complacency, the first phase of what hydrologists call a "hydro-illogical cycle" that characterizes government's traditional response to drought. Even though drought is as inevitable, if not predictable, as the change of seasons, governments tend to ignore the danger when water is abundant. As a result, they are slow to see it coming and ill-prepared for it when it arrives. Then, when they belatedly realize they are in a drought, their responses often are either ineffectual or even counterproductive.

"In retrospect, we should have been watching soil moisture and tree moisture, not just reservoir levels and snow pack," says Elizabeth Gardner, Denver Water's manager of conservation. "We also needed to make people recognize the seriousness of the drought earlier. And we needed a drought plan that was much more detailed."

In the hydro-illogical cycle, concern can quickly turn to panic in the absence of comprehensive drought planning. Unfortunately, panic rarely makes for good policy. In Colorado, for instance, drought has led officials to take a new look at sweeping proposals that had long been rejected as too costly or inimical to Coloradoans' sense of their own state. One proposal, dubbed the "Big Straw," would involve pumping water from the Colorado River near the Utah border back to the Continental Divide. The cost would be staggering: The state would have to build a 200-mile pipeline, at least two large reservoirs, a pumping system that would require the equivalent of 80 percent of the annual output of Hoover Dam to lift the water 4,000 feet over the Rockies, and new systems for cooling and purifying the water—all at a potential cost of $5 billion.

An even more draconian idea would involve clear-cutting large swaths of federal and state forests so that more snow would reach the ground, where, the theory goes, it eventually would melt and run into the streams that supply city reservoirs. "Logging for water," as that idea is known, could require cutting as much as 40 percent of the trees in watersheds, a process that not only would be expensive but also could damage natural habitats and do untold harm to the recreation and tourism industries.

Critics say such costs are not only destructive but also entirely avoidable. With prudent planning, they contend, a state such as Colorado can manage droughts for many years without taking such drastic and costly steps. "The cities of Colorado don't have to worry about running out of water for decades, or even centuries," says Douglas Kenney, research associate at the University of Colorado's Natural Law Resources Center. But, he adds, avoiding panic-induced measures will require policy makers to develop a whole new mind-set. "We have to stop thinking about drought as a phenomenon to be avoided at all costs, and think of it instead as a normal part of life," Kenney argues. "We have to get more used to the idea of risk management."

Fighting back

Risk management could be the byword of a new drought strategy that is taking shape in Georgia, where a broad-based working group has proposed a state drought plan that emphasizes permanent reforms rather than emergency responses. "We decided that a drought plan isn't about what you do in a drought, it's about what you do in advance to mitigate the effects of drought," says Robert Kerr, who, as director of the Pollution Prevention Assistance Division for the state Environmental Protection Division, led the planning effort.

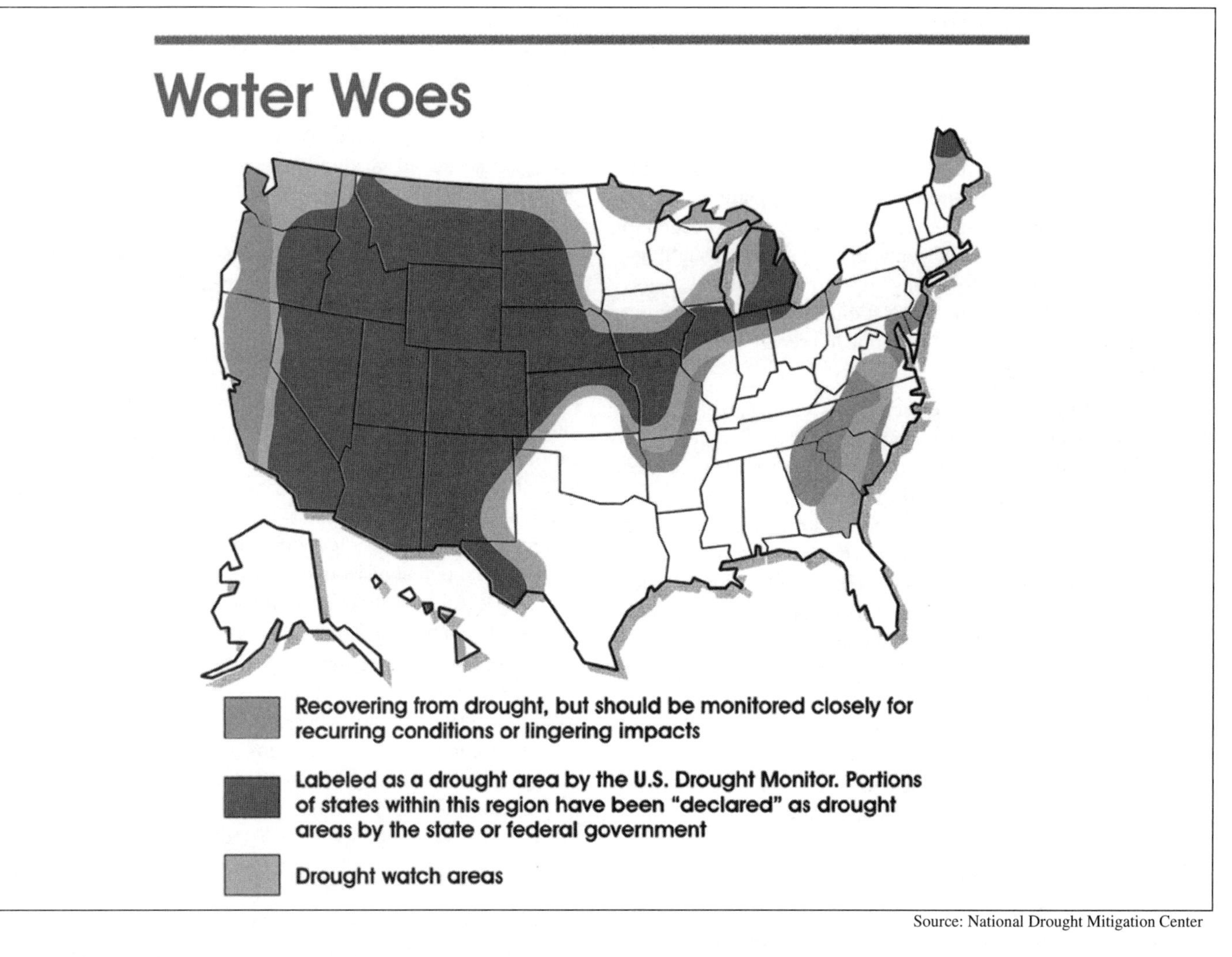

Source: National Drought Mitigation Center

One key proposal, for instance, would regulate outdoor watering all the time—not just during droughts. The plan is controversial even though it would allow homeowners to water their yards three times a week, enough to keep any home landscape in good health. Indeed, it would make many yards healthier than they are today. That's because most homeowners tend to overwater their plants, a practice that prevents the plants from developing deep root systems that enable them to find water if the soil around them dries up. Rose Mary Seymour, an engineer and horticulturist at the University of Georgia, argues that modest restrictions on watering would both reduce the risk of drought by saving water and make droughts that do occur less damaging. Seymour works with landscapers and gardeners to encourage xeriscape, a system of water-efficient landscaping design and maintenance.

The focus on outdoor watering reflects a simple fact: For many municipal water systems, landscape irrigation represents the biggest, and probably least essential, category of water consumption. In Denver, for instance, outdoor watering accounts for 40 percent of all water use. But Georgia's growing commitment to water conservation involves other sectors as well. The state has been conducting water audits of major institutions, and the drought-planning group has studied ways to improve the efficiency of agricultural irrigation. In addition, the Metropolitan North Georgia Water Planning District, a state-created agency that charts water policy for municipal governments in the Atlanta region, is preparing a list of urban water conservation measures it believes could reduce water demand by as much as 15 percent—or 1 billion gallons a day—by 2030.

Among the most promising reforms: requiring that existing homes be retrofitted with newer, water-efficient shower heads and toilets; paying homeowners rebates for buying more efficient washing machines; mandating rain-sensors that turn off automatic watering systems when they aren't needed; requiring the use of water meters for multi-family as well as single-family homes; and doing water audits for homes, hotels and commercial establishments.

Of course, even aggressive conservation won't avert all droughts, Georgia officials concede. So their plan also calls for a sophisticated early warning system. Instead of relying on a single drought indicator, as Denver Water officials did last year, the Georgia plan will look at four indicators—stream flows, groundwater levels, reservoir levels and precipitation patterns. Anytime one or more of these indicators drops below a certain level in a particular climate zone, the state would order new restrictions on outdoor watering. And those limits would stay in effect until all of the "triggers" that indicate drought conditions have taken a turn for the better.

Authority figure

Georgia's proposals represent a significant increase in the role of state government, but officials believe they are necessary to help avoid problems such as those in Denver last summer. There, a plethora of independent municipal water systems all decided for themselves whether there was a drought, how serious it was and what should be done about it. Officials at Denver Water believe the mixed messages they sent the public are a major reason its own watering restrictions proved ineffective for six crucial weeks last summer. (Denver-area water managers subsequently began meeting to develop a more coordinated effort.) Georgia officials believe their approach is the right balance between state and local control: The state would set minimum requirements and municipalities would be free to adopt more stringent standards. "There needs to be a state coordinating mechanism, but it needs to be sensitive to the local context," says Anne Steinemann, a Georgia Institute of Technology drought expert who advised Georgia on its planning effort.

Taking Drought's Measure

The annual cost of drought:

$ 6 billion to $8 billion

The annual cost of flooding:

$ 5.9 billion

The annual cost of hurricanes:

$5.1 billion

Source: National Weather Service

The Georgia plan, however, also leaves some of the toughest decisions to local officials. For one thing, it gives them the responsibility to decide whether to reform water rates to promote conservation. Currently, many municipal water systems charge a flat rate regardless of how much water is consumed. Some even offer lower rates to heavy consumers. The lack of a clear price signal is one reason why some homeowners "turn their yards into rice paddies," says Roy Fowler, general manager of the Cobb County-Marietta Water Authority, a wholesale supplier for municipal water systems in the Atlanta area. Cobb-Marietta imposes a 25 percent surcharge on customers whose summer consumption exceeds 130 percent of their winter consumption. That, combined with such conservation measures as federally mandated water-efficient toilets, helped reduce per capita consumption among its customers from 146 gallons a day in 1990 to 130 gallons in 2000.

Fowler is one of a new breed of water managers. Traditionally, water managers defined their jobs mainly in terms of supply—that is, they sought to provide as much water as consumers wanted, with no questions asked. But today, Fowler says, "managing demand" is just as important. The reason boils down to basic economics: Conservation saves money. "A conscientious conservation program," Fowler notes, "costs pennies on the dollar compared to digging a hole in the ground and calling it a reservoir."

Still, conservation creates a new kind of risk for water managers. As they wring the inefficiencies out of current urban water systems, they will have fewer options for finding easy savings in the future. Aware of this problem, they are starting to look for other ways to minimize the disruption they could face in the worst droughts. Robert Kerr, who is shepherding the Georgia plan through the regulatory process, believes the ultimate solution may be the water equivalent of "rolling blackouts," in which some industries might be willing to curtail operations on certain days in return for a price break on their water.

A shared solution

Long before that happens, cities could negotiate arrangements to tap into water generally used for agricultural irrigation during serious droughts. That represents a substantial potential reserve in most states. In Georgia, for instance, farm irrigation accounts for 30 percent of the water consumed during non-drought years. What's more, there are ample models for how such a water-insurance system could work: Federal farm policies have made farmers well accustomed to the idea of forgoing production in exchange for cash payments. Georgia already has applied the principle to water. In 2001 and 2002, both serious drought years, the state conducted auctions to buy back irrigation rights from farmers. Last year, the auctions took 42,000 acres out of irrigation at a cost of $5.5 million, an arrangement that added about 25 percent to the flow of the Flint River in the southern part of the state.

Water-sharing arrangements have even more potential in the arid West, where agriculture accounts for a larger percentage of total water consumption than in Eastern states such as Georgia. Indeed, some Colorado municipalities already have negotiated deals to lease farmers' water rights during emergencies, and environmental groups in the Rocky Mountain state are pushing legislation that would clear potential legal barriers to additional transfers. Environmentalists see the idea not only as an alternative to mega-projects like the "Big Straw" but also as a way to preserve open space. There's no question that cities have the economic clout to buy up all the rural water they need for the foreseeable future, the environmentalists note. But cities will buy less—and hence keep more land in agricultural production longer—if they can reach agreements that ensure the water will be available to them in emergencies.

Such stand-by arrangements may sound like a common-sense solution, but they aren't a sure thing politically. In Georgia, water auctions have proven to be controversial with city voters. "A lot of people say, 'I had to cut off watering my lawn and I didn't get paid. Why should farmers?'" notes Jim

Hook, a soil and water management specialist at the University of Georgia. "Farmers are such a small political constituency it's not clear there will be the political will to keep the water auctions up."

That points to perhaps the biggest challenge of all when it comes to planning for drought. Water issues are, to say the least, politically divisive. Many leaders are reluctant to take them on when there is no crisis to force the issue. Still, you don't have to be a weatherman to see that the issue can't be avoided forever. With continued population growth, conditions that are considered drought today could be the norm soon. Denver Water's long-range plans show, for instance, that the same degree of conservation the city ultimately achieved on an emergency basis last year—by fall, it had cut water consumption by 30 percent—will have to be a permanent way of life by 2050.

Will Denverites take the goal seriously once the snows return and refill their reservoirs? The eventual success of water restrictions last year offers reason to believe they will. But the hydro-illogical cycle suggests another outcome: a return to apathy. As John Steinbeck, chronicler of America's "Dust Bowl" drought years, noted, "It never failed that during the dry years people forgot about the rich years, and during the wet years they lost all memory of the dry years. It was always that way."

An Indian Paradox: Bumper Harvests And Rising Hunger

The World Has Enough Food, But Poor Can't Afford It; Growing Jobs and Crops

A Taxi For Mr. Managatti

ROGER THUROW AND JAY SOLOMON

THIRUKANCHIPET, India—In the 1960s, this country set out to prevent famine by boosting agricultural production. The push was so successful that wheat and rice stockpiles approached 60 million tons. By 2001, India had its own grain export business. But Murugesan Manangatti, a 29-year-old illiterate peasant, was still hungry. He had no land to grow crops and no steady income to buy food.

Last summer, an agricultural research foundation gave Mr. Manangatti some unusual advice: Drive a taxi. With the foundation's help, he and 15 members of this rural village received a loan to buy a three-wheeled, battery-powered vehicle. The taxi business earns up to $25 a day and Mr. Manangatti takes home a monthly salary of about $55. For the first time, he says, his family is regularly able to eat three nutritious meals a day.

The Thirukanchipet taxi is a fresh approach to solving a jarring paradox. The world is producing more food than ever before as countries such as India, China and Brazil emerge as forces in global agriculture. But at the same time, the number of the world's hungry is on the rise—including in India—after falling for decades. Despite its overflowing granaries, India has more hungry people than any other country, as many as 214 million according to United Nations estimates, or one-fifth of its population.

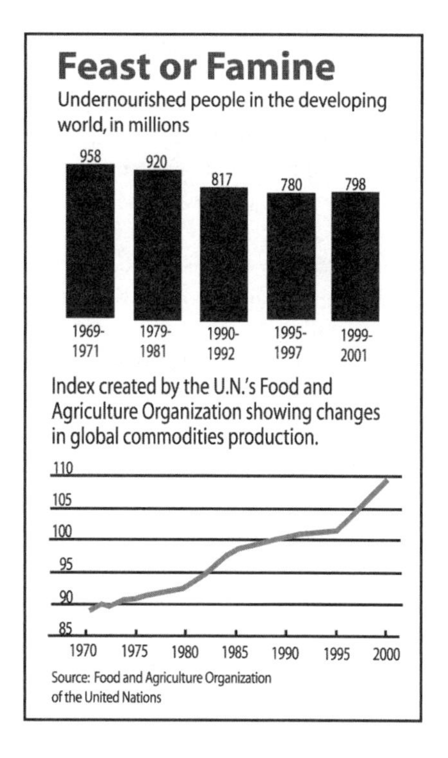

The paradox is propelling a shift in strategy among the world's hunger fighters. International agencies that once encouraged countries to solve starvation crises by growing more food are now tackling the more fundamental problem of rural poverty as well. The old development mantra—produce more food, feed more people—is giving way to a new call: Create more jobs, provide income to buy food.

"Increasing production is great, but we have to think about the whole chain," says M.S. Swaminathan, the 78-year-old scientist who helped engineer India's agriculture boom and whose foundation set up Mr. Manangatti's taxi. India has been able to conquer its famine of food, he says. Now it is suffering from a "famine of jobs and livelihoods."

The stark contrast between food production and rural poverty is helping to transform Indian politics. India's ruling Bharatiya Janata Party had overseen a boom in the country's technology sector but was defeated in May elections largely by the votes of a rural population that felt left behind. The BJP's "India Shining" campaign, which highlighted the country's economic advances, was trumped by the victorious Congress party, which ran on a platform of aiding farmers.

There is plenty of supply on hand to meet global demand. Over the past 35 years, the world's food production has expanded faster than its population. In 2002, according to the United Nations World

Food Program, farmers produced enough food to provide every person with 2,800 calories a day. That's equivalent to the general daily requirement of teen boys and active men, according to the U.S. government's dietary guidelines. The WFP's feeding programs aim to provide 2,100 calories a day to their recipients.

But inadequate infrastructure, local corruption and rural poverty have prevented the chronically hungry—those who don't eat enough to fulfill basic standards—from gaining access to this bountiful harvest. After falling for decades, the estimated number of undernourished in the developing world increased by 18 million to 798 million between 1997 and 2001, according to the latest data from the U.N.'s Food and Agriculture Organization.

In a typical year, the World Food Program distributes food to about 90 million people, many of whom are threatened with starvation in disaster situations such as drought. Most of the remaining 700 million live on isolated, stingy land, and have neither the money to buy food nor the ability to grow it. They're beyond the reach of international feeding programs and also fall through national safety nets.

It's virtually impossible to simply hand out food surpluses to the hungry because of the cost and complexity of distribution. It would also turn recipients into permanent wards of the world. "I believe in Gandhi's strategy: Don't turn people into beggars," says Mr. Swaminathan.

Looking for solutions, countries are turning their attention to permanent development projects such as road building that can foster economic activity for the rural poor, and connect them to markets for their produce.

A recent summit meeting of the Group of Eight industrialized nations embraced a plan to "end the cycle of famine" in the Horn of Africa. One plan for Ethiopia involves creating work programs that would allow the five million people there dependent on aid to buy their own food. Earlier this year, the Chinese government said it would cut farming taxes and boost investment in rural areas. And at a recent meeting of the African Union, leaders committed to allocating at least 10% of their budgets to agriculture and rural development.

The WFP, in one strategy shift, is emphasizing schools with a classroom-based feeding program that so far reaches 15 million. It's designed to encourage children, who constitute about 300 million of the world's hungry, to attend school and at the same time combat malnutrition. "An ill-educated, unhealthy population can't take advantage of an open economy," says John Powell, a WFP deputy executive director.

India's agricultural program of the 1960s, dubbed the Green Revolution, was launched after the country suffered through a series of famines. Under the guidance of local and international agronomists and scientists, Indian farmers were introduced to hardy, fast-growing wheat strains and better uses of fertilizer and irrigation. As a result, crop yields multiplied and in recent years India's wheat production topped 70 million tons, surpassing that of the U.S. The Indian government estimates that wheat output may pass 100 million tons in the coming decade.

In the country's northern grain belt, wheat grows almost everywhere there is a level field, between houses and schools, and brick factories and gas stations. During the harvest season, the roads are clogged with tractors such as the small Massey Ferguson model driven by 19-year-old farmer Gopal Kumar. He recently pulled a wagon piled with five tons of wheat as he made his way to the mill in Mathura, a 30-minute drive from the Taj Mahal.

The Kumars have been farming wheat for three generations and now work 36 acres. "This is a pretty good year," said Mr. Kumar. He maneuvered his tractor and wagon past Mathura's McDonald's and delivered his wheat to Rajender Bansal's mill. Mr. Bansal opened his mill about 10 years ago with capacity to process 2,000 tons of wheat a month. He has since expanded to 3,000 tons.

As India's grain production grew, so did its surpluses. By 2001, the national stockpile of rice and wheat was approaching 60 million tons, according to the government. The country had also become one of the world's leading producers of fruits, vegetables and milk. India set up a distribution network to supply surplus grain at reduced prices to 180 million families.

But with inefficiency and local mismanagement plaguing distribution, it couldn't move the grain fast enough through the system. Some even spoiled in warehouses. A 2002 government survey concluded that 48% of children under five years old are malnourished. That's an improvement from three decades ago and even today, given rapid population growth, the proportion of chronically hungry Indians continues to fall. But in a sign that there are limits to the Green Revolution, the absolute number of hungry people in India began to rise again in the late 1990s, according to the U.N.

With the cost of storing surpluses spiraling, the government opened the door to grain exports in 2001. India sold more than 10 million tons of grain to overseas customers that year, mostly in Asia and the Middle East.

Traders from traditional wheat and rice exporters were critical of the Indian trade. How could the country export grain while so many in the country are hungry? D.P. Singh, chairman of the All India Grain Exporters Association, says the grain surplus has been big enough to allow for both exporting and distribution to the rural poor. "If [the grain] didn't reach the hungry people, it's too bad, but it has nothing to do with availability," he says.

At the same time, India made a donation of one million tons of wheat to a World Food Program project in Afghanistan. A few European members of the WFP's executive board questioned the propriety of India's action. Himachal Som, India's representative to the U.N.'s food agencies in Rome, made an impassioned speech to his critics arguing that the donation didn't affect the country's ability to feed its poor, a more intractable problem than simply growing greater amounts of food.

The results of last month's election in India are concentrating attention on the paradox of hunger. In the two states where the former BJP-led government fared especially badly—Andhra Pradesh and Tamil Nadu—the gap between India's high-tech centers and surrounding farming areas had become the most pronounced. Hyderabad, the capital of Andhra Pradesh, grew prosperous as the state's government courted U.S. companies such as Microsoft Corp. and General Electric Co. and the World Bank praised the state for its economic progress.

But about 100 miles outside the city's glittering office towers, farmers in the town of Kalimela say they've benefited little. A three-year drought hit farm production. Many blamed the state government for failing to invest more in irrigation systems and roads. In addition, farmers were hit hard when the state increased electricity rates.

"The government hasn't helped us. No roads. No water. Right from the beginning," says Jarappa Sonia, 35, a sugar cane and wheat farmer from Kalimela. Mr. Sonia joined a government work program building roads in a district four hours from his home. "I prefer to be a farmer," he says.

The opposition Congress party promised free power for Andhra Pradesh's farmers. Within hours of taking office, the state's new chief minister, Y.S. Rajasekhar Reddy, honored that pledge.

It's too early to know how fully the Congress-led government will implement its ideas. The Congress party and its allies have agreed to support minimum-wage public-works programs such as a guaranteed 100 days of employment for rural households. They have also promised to improve farmers' access to credit and restructure outstanding debts.

Providing rural folk with an income to buy food is a theory the Swaminathan foundation has extended to 9,600 people in 800 self-help groups in five states across the country.

In the small southern village of Thirukanchipet, the best the rural unemployed can hope for is seasonal work in rice paddies for $1 or $2 a day. Earlier this year, with the help of the Swaminathan foundation, 28 men and women formed a dairy group to improve their credit worthiness and received a micro-credit loan of about $10,000 from a local bank.

The group bought 20 cows and 19 calves, and built a milking shed. The cows produce more than 40 gallons a day, which is sold to a local dairy cooperative for 80 cents a gallon. Much of the daily income of about $30 is set aside to repay the loan. The rest is distributed among the members who for the first time are able to afford the higher-quality rice and wheat sold in the private stores instead of that in government ration shops.

"It tastes better," says M. Kanagaraj, a tall, thin man of 34. He is one of four workers who milk and manage the cows and makes about $40 a month. "We eat two meals a day now," he says. They are waiting for the calves to mature, so they can double the milk production and their income.

"Then," says Mr. Kanagaraj, "we will eat three meals a day."

Test Your Knowledge Form

We encourage you to photocopy and use this page as a tool to assess how the articles in *Annual Editions* expand on the information in your textbook. By reflecting on the articles you will gain enhanced text information. You can also access this useful form on a product's book support Web site at *http://www.dushkin.com/online/*.

NAME: DATE:

TITLE AND NUMBER OF ARTICLE:

BRIEFLY STATE THE MAIN IDEA OF THIS ARTICLE:

LIST THREE IMPORTANT FACTS THAT THE AUTHOR USES TO SUPPORT THE MAIN IDEA:

WHAT INFORMATION OR IDEAS DISCUSSED IN THIS ARTICLE ARE ALSO DISCUSSED IN YOUR TEXTBOOK OR OTHER READINGS THAT YOU HAVE DONE? LIST THE TEXTBOOK CHAPTERS AND PAGE NUMBERS:

LIST ANY EXAMPLES OF BIAS OR FAULTY REASONING THAT YOU FOUND IN THE ARTICLE:

LIST ANY NEW TERMS/CONCEPTS THAT WERE DISCUSSED IN THE ARTICLE, AND WRITE A SHORT DEFINITION:

We Want Your Advice

ANNUAL EDITIONS revisions depend on two major opinion sources: one is our Advisory Board, listed in the front of this volume, which works with us in scanning the thousands of articles published in the public press each year; the other is you—the person actually using the book. Please help us and the users of the next edition by completing the prepaid article rating form on this page and returning it to us. Thank you for your help!

ANNUAL EDITIONS: Geography 05/06

ARTICLE RATING FORM

Here is an opportunity for you to have direct input into the next revision of this volume.
We would like you to rate each of the articles listed below, using the following scale:

1. **Excellent: should definitely be retained**
2. **Above average: should probably be retained**
3. **Below average: should probably be deleted**
4. **Poor: should definitely be deleted**

Your ratings will play a vital part in the next revision.
Please mail this prepaid form to us as soon as possible.
Thanks for your help!

RATING	ARTICLE
___________	1. The Big Questions in Geography
___________	2. Rediscovering the Importance of Geography
___________	3. The Four Traditions of Geography
___________	4. The Changing Landscape of Fear
___________	5. Recreating Secure Spaces
___________	6. Perilous Gardens, Persistent Dreams
___________	7. After Apartheid
___________	8. How Cities Make Their Own Weather
___________	9. The Race to Save a Rainforest
___________	10. Texas and Water: Pay Up or Dry Up
___________	11. Environmental Enemy No. 1
___________	12. Carbon Sequestration: Fired Up With Ideas
___________	13. Trading for Clean Water
___________	14. Every State is a Coastal State
___________	15. The Rise of India
___________	16. Between the Mountains
___________	17. A Dragon With Core Values
___________	18. L.A. Area Wonders Where to Grow
___________	19. Reinventing a River
___________	20. Unscrambling the City
___________	21. An Inner-City Renaissance
___________	22. On the Road to Agricultural Self-Sufficiency
___________	23. Mapping Opportunities
___________	24. Geospatial Asset Management Solutions
___________	25. The Future of Imagery and GIS
___________	26. Internet GIS: Power to the People!
___________	27. ORNL and the Geographic Information Systems Revolution
___________	28. Europe's First Space Weather Think Tank
___________	29. Mapping the Nature of Diversity
___________	30. Fortune Teller
___________	31. Resegregation's Aftermath
___________	32. A City of 2 Million Without a Map
___________	33. The Longest Journey
___________	34. Borders Beyond Control
___________	35. China's Secret Plague
___________	36. The Next Oil Frontier
___________	37. Mexico: Was NAFTA Worth It?
___________	38. Dry Spell
___________	39. An Indian Paradox: Bumper Harvests and Rising Hunger

(Continued on next page)

NO POSTAGE
NECESSARY
IF MAILED
IN THE
UNITED STATES

BUSINESS REPLY MAIL
FIRST CLASS MAIL PERMIT NO. 551 DUBUQUE IA

POSTAGE WILL BE PAID BY ADDRESEE

McGraw-Hill/Dushkin
2460 KERPER BLVD
DUBUQUE, IA 52001-9902

ABOUT YOU

Name — Date

Are you a teacher? ❒ A student? ❒
Your school's name

Department

Address — City — State — Zip

School telephone #

YOUR COMMENTS ARE IMPORTANT TO US!

Please fill in the following information:
For which course did you use this book?

Did you use a text with this ANNUAL EDITION? ❒ yes ❒ no
What was the title of the text?

What are your general reactions to the *Annual Editions* concept?

Have you read any pertinent articles recently that you think should be included in the next edition? Explain.

Are there any articles that you feel should be replaced in the next edition? Why?

Are there any World Wide Web sites that you feel should be included in the next edition? Please annotate.

May we contact you for editorial input? ❒ yes ❒ no
May we quote your comments? ❒ yes ❒ no